SMALL-SCALE GOLD MINING

A MANUAL BASED ON EXPERIENCE IN SURINAME

BY DR. E. H. DAHLBERG

Practical
ACTION
PUBLISHING

Intermediate Technology Publications 1984

Practical Action Publishing Ltd

25 Albert Street, Rugby, CV21 2SD, Warwickshire, UK

www.practicalactionpublishing.com

© Intermediate Technology Publications 1984

First published 1984\Digitised 2013
Reprinted in the UK, 2020
Reprinted by Practical Action Publishing
Rugby, Warwickshire UK

ISBN 10: 1 85339 049 6
ISBN 13: 9781853390494
ISBN Library Ebook: 9781780443348
Book DOI: http://dx.doi.org/10.3362/9781780443348

A catalogue record for this book is available from the British Library.

Since 1974, Practical Action Publishing has published and disseminated books and information in support of international development work throughout the world. Practical Action Publishing is a trading name of Practical Action Publishing Ltd (Company Reg. No. 1159018), the wholly owned publishing company of Practical Action. Practical Action Publishing trades only in support of its parent charity objectives and any profits are covenanted back to Practical Action

FOREWORD

For some years now it has been the endeavour of the Intermediate Technology Development Group to assist small scale mining in the belief that it holds great potential for employment generation. This sector is relatively neglected by the technical development agencies, and it is very clear to us that improved, simple techniques for prospecting, mining and processing are a major priority for advancement.

When, therefore, the Manual of Gold Mining was sent to us from Suriname, we felt that the information it contained would be useful in many other countries.

Gold mining is one of the few operations that has continued to be successfully practised in spite of world recession and low metal prices. This has meant that it was taken up by many people with little knowledge of the techniques required and has sometimes led to inefficient and sub-economic efforts. It is the purpose of this manual to show firstly how systematic prospecting can ensure better grades and a longer life of the deposit and, secondly, how simple equipment, which can be made locally, can reduce the effort required and at the same time considerably improve the recovery of the gold. It deals only with alluvial operations but there are so many locations all over the world where this is the first introduction to mining activities, that we are certain that its simple, practical approach will make a fundamental contribution in many areas.

Several of the chapters, especially those dealing with mining regulations, are specific to Suriname, but we have not omitted them as it is always interesting to compare approaches in other countries. One of the methods described uses mercury and we have therefore added a chapter on mercury poisoning. We have also added a short list of references for those who wish to go deeper into any part of the subject.

Our thanks are due to Dr. E. H. Dahlberg for allowing us to translate and publish this volume.

ITDG, Rugby,
January, 1984

CONTENTS   <u>Page</u>

1.   Introduction   1

2.   Application for a Concession   3
            and Duties of the applicant

3.   Types of gold deposits in   5
            Suriname

4.   Exploration for Gold   12

    4.1   Gold Concentrating Methods   12

    4.2   Prospecting   27

    4.3   Exploration   31

5.   Purity of gold   44

6.   Amalgamation of gold from   44
            mineral concentrates with
            mercury

7.   Treatment of poisoning   46

8.   References for further reading   49

1.   <u>INTRODUCTION</u>

As a result of the increased price of gold,
interest in gold mining in Suriname is growing
once again. This development is encouraged by
the GMD, because their primary function is to
support mining activities. This manual was
written for people who want to take up gold
mining, and as such is part of the facilities
provided by the GMD. The saying 'look before you
leap' is more relevant here than in any other
undertaking. Gold's lucrative character,
interwoven with legendary stories of rich gold
occurrences, has in the past resulted in many
hasty mining ventures, doomed to failure.

The technical and commercial aspects must be very
carefully thought through. In the following
pages we shall try to acquaint the interested
outsider with the nature of gold deposits and
with factors which have to be borne in mind
before starting out on a mining venture.

Suriname had a prosperous gold mining industry at
the turn of the century on a large, medium and
small scale. The collapse of that industry was
caused by:

-    the exhaustion of easily exploitable
     areas and a lack of prospecting and
     exploration activities to establish
     further reserves.

-    lack of management expertise resulting
     in unsound mining plans and purchases
     of unsuitable equipment.

-    the waste during mining of gold
     occurring in the form of fine
     particles resulting in substantial
     losses.

-    poor control of selling of gold
     resulting in the formation of a black
     market and low prices for the
     producer.

-    sub-leasing by the concession holder
     to small producers resulting in an
     unsatisfactory relationship.

- the freezing of the gold price on the world market.

This handbook will show, in logical order, the steps necessary to bring an economically exploitable gold deposit into production.

2. APPLICATION FOR A CONCESSION AND DUTIES OF THE
   APPLICANT

   The G.M.D. (Geological and Mining Service)

   Before embarking on an investigation it would be
   advisable to contact the head of the GMD to
   obtain information on the following:-

       (a) a list of areas which are open to
           prospecting

       (b) a list of areas which are likely to be
           promising from a geological and mining
           point of view.

   Regarding (a) the GMD can issue a map which shows
   all existing concessions and their expiry dates
   as well as areas which have been reserved by the
   government and which may not be traversed by
   concession holders.

   Regarding (b) the following criteria will apply:-

       -   the possibility of gold occurrence
           based on geological data

       -   the possible occurrence of gold based
           on verbal or written information

       -   occurrence of gold established by
           prospecting

       -   reserves of gold established by
           exploration

   The annual cost of an exploration licence is 1
   cent per hectare and for exploitation 10 cents
   per hectare.* The Mining Regulations limit the
   maximum size of concessions to 20,000 hectares.

---

* 1 Suriname Guilder = 100 cents = £4.5 (approx.)

At the end of every 3 months the concessionaire
has to present a report to the head of the GMD
with the following details:-

- list of people employed and their
  addresses

- the salaries paid to them

- technical details giving the number of
  pits dug, the amount of ground moved,
  the weight of gold found, and other
  information in accordance with
  paragraphs 4.2.2 and 4.3.1 below.

The report must be accompanied by a map giving
the pit and sample locations. It must also be
pointed out that the gold found by an exploration
licence holder is not his property, but must be
handed over to the head of the GMD.

Office of Lands Administration

After consultation with the GMD one or more areas
are chosen, and a decision is made to apply for
licences. The procedure is set out in Government
Pamphlet (GB) 1952 No. 28 and 29 of the Mining
Regulations. It will be necessary to consult a
chartered surveyor who will advise whether any
particular area can be applied for. The surveyor
or applicant then sends the application to the
Office of Lands Administration (a department of
the Ministry of Development) and approval usually
follows in 6 to 7 weeks. Before this approval,
the head of the GMD can, by letter, give
permission for exploration to start. If the
geological and mining information is
satisfactory, then an exploitation licence can be
applied for at the same time as the exploration
licence. In the case of a navigable stream or
river, it is recommended to apply immediately for
an exploitation licence in view of the
complicated aspects of exploration, and so that
favourable results can immediately be followed by
mining.

3.   TYPES OF GOLD DEPOSITS IN SURINAME

Gold occurs in Suriname in the following ways:-

A.   As a primary deposit in crystalline
rocks, also called bedrock. This gold
occurs either in or in the immediate
vicinity of quartz veins (reefs).

B.   As a secondary deposit in the vicinity
of gold bearing eroded bedrock. These
are called colluvial deposits.

C.   As a secondary deposit in recent and
abandoned stream and river beds which
drain the water from colluvial
deposits. They are called alluvial,
stream or river deposits and occur in
the areas of large streams and rivers.

The deposits under B and C are also called placer
deposits.

A - primary gold deposits

In Suriname these occur on a regional scale at
contact zones of quartz diorite and metamorphic
volcanic rock which form the matrix or bedrock,
see Figs. 1 and 2. Gold occurs mainly in the
immediate surroundings of the quartz veins and
adjoining rock as shown in Figs. 3 and 4. The
quartz veins can be milky white or grey to blue.
The museum of the GMD has a manual illustrating
the various rock formations.

B - secondary colluvial deposits

These deposits occur in the immediate vicinity of
gold bearing bedrock, at the foot of the slope
along the stream bed. They were formed by the
disintegration and selective transport of the
primary rock over short distances by gravity and
flowing surface water. They consist mainly of
yellow-brown to red-brown lumps of clay
containing iron nuggets and sharp pieces of
quartz, as shown on Fig. 5.

C - secondary alluvial deposits

These occur in recent or abandoned stream beds
and adjoining terraces which drain areas where
gold-containing bedrock and secondary colluvial
deposits occur.

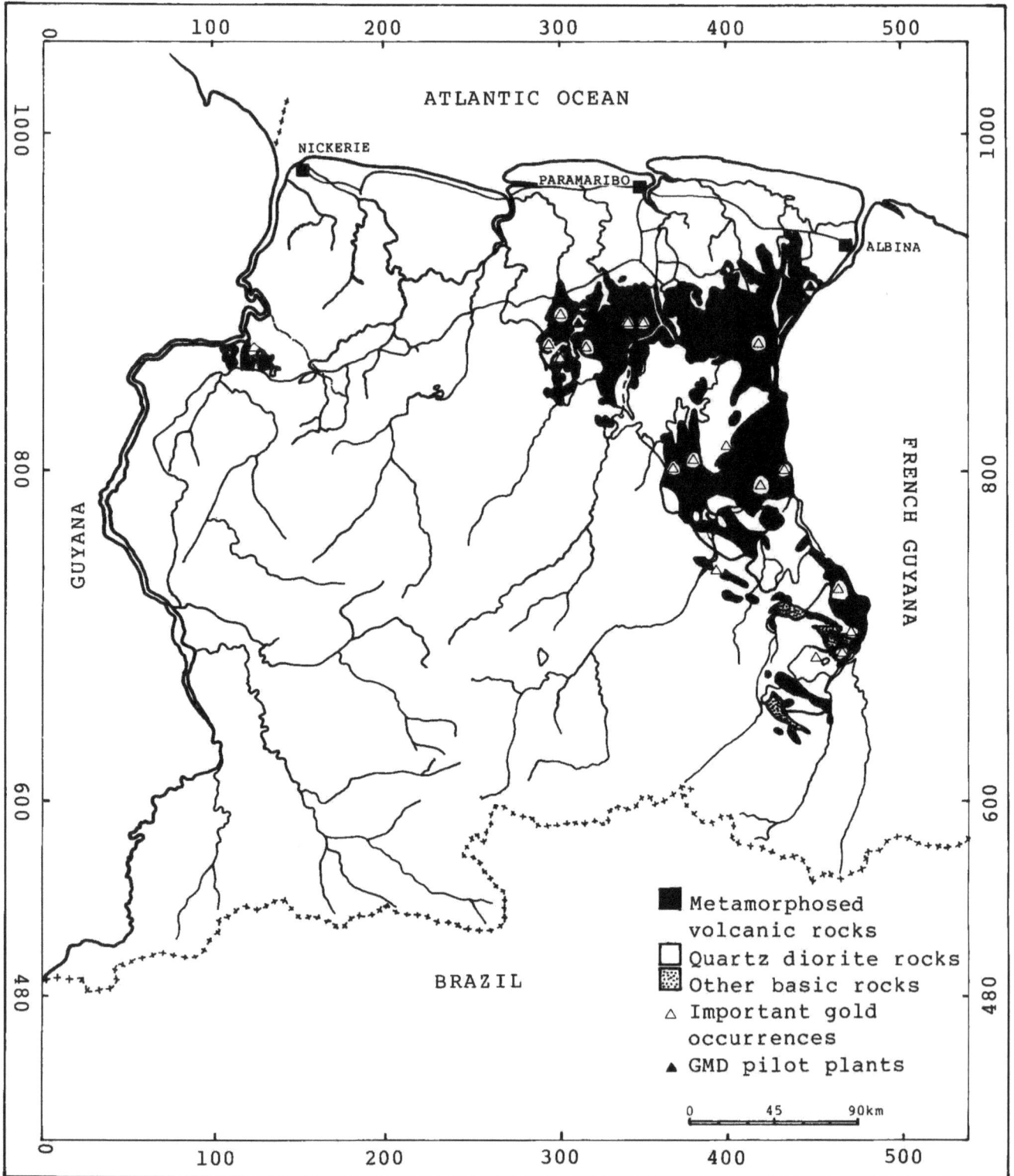

Fig. 1 Map of Suriname showing important gold occurrences and gold bearing formations

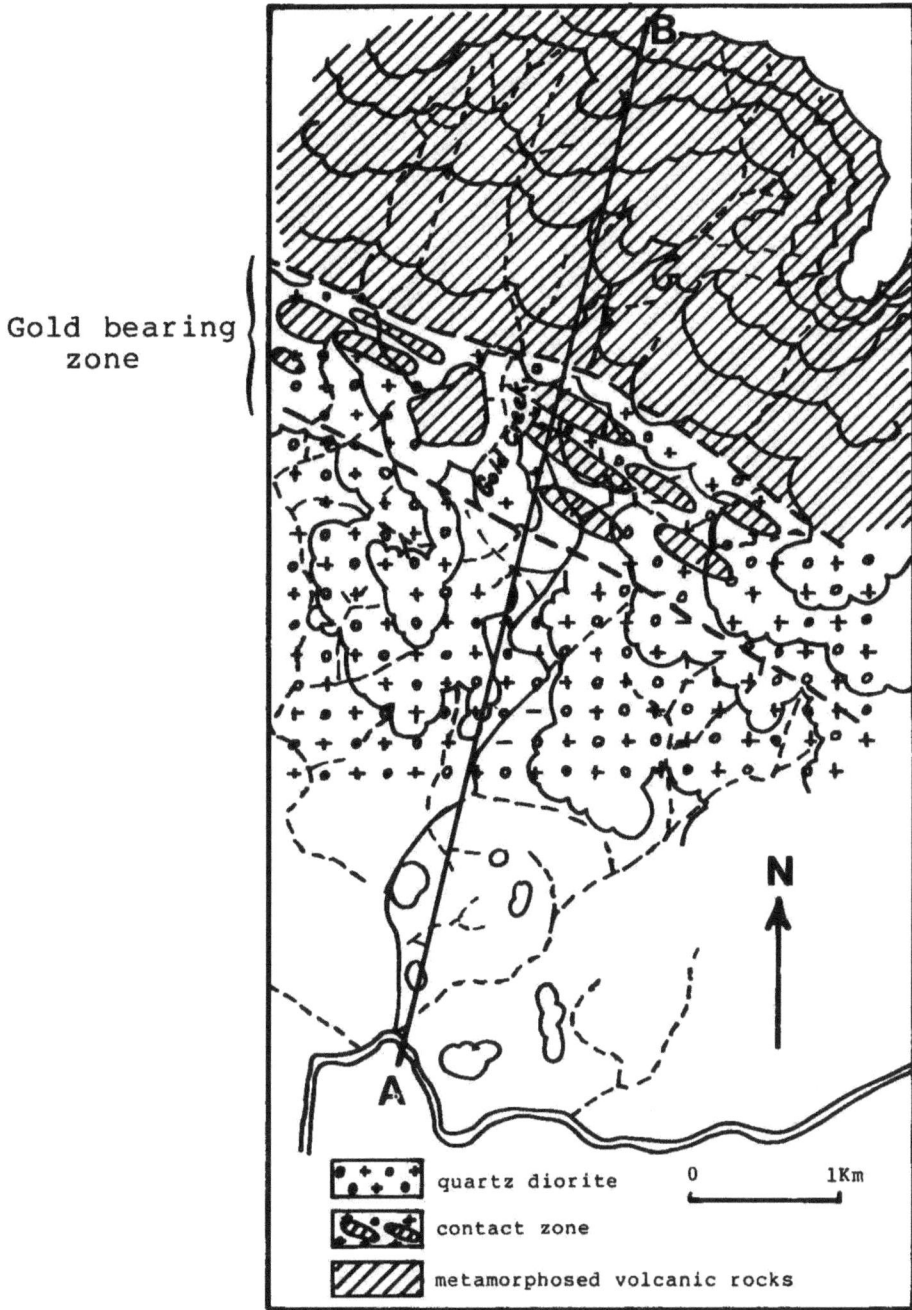

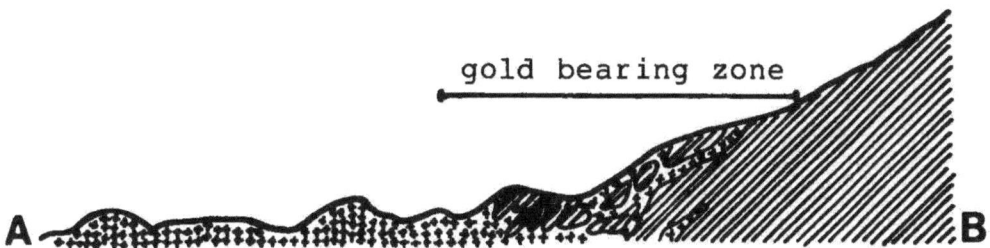

Fig. 2 Map and section with gold bearing contact
zone between volcanic and dioritic rocks
(According to Brinck, 1955)

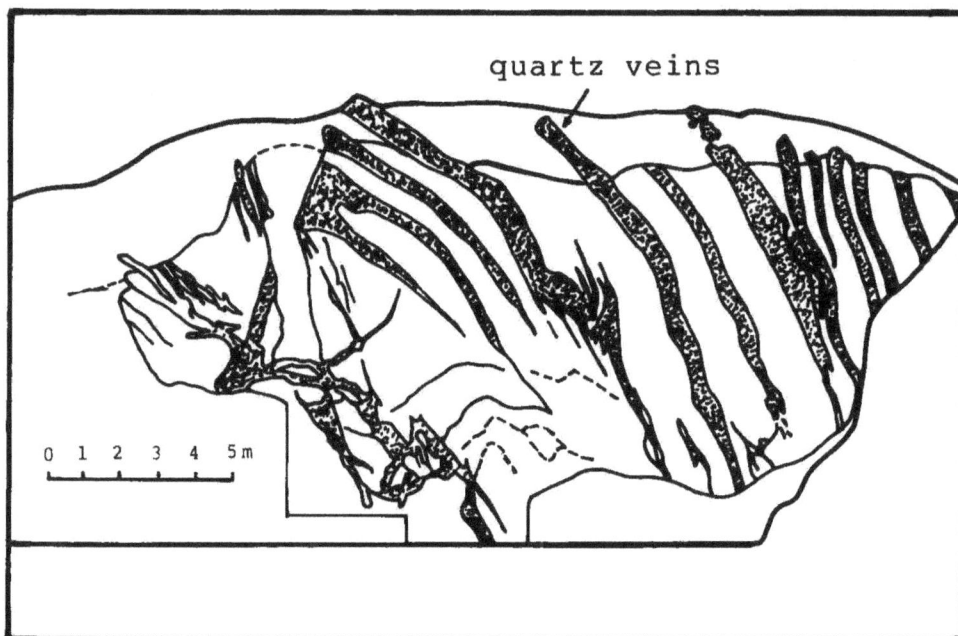

Fig. 3   Section with a series of quartz veins in
weathered rock which can contain gold.
Suitable for investigation with prospecting
pits.

red-yellow weathered soil, sometimes with iron bands

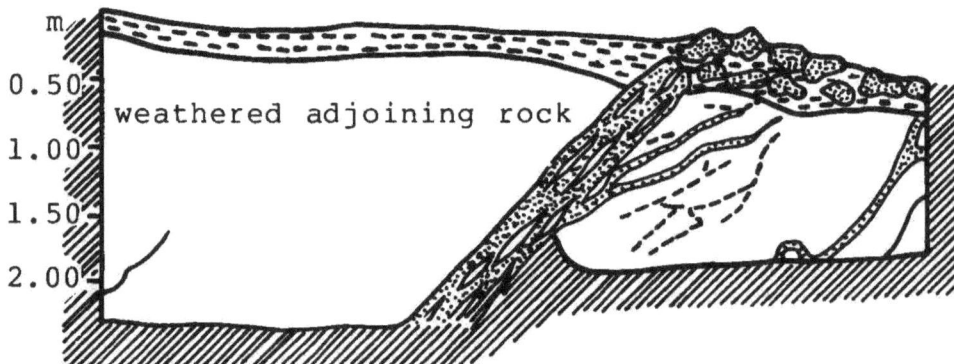

Fig. 4   Quartz veins which downhill become colluvial
deposits with angular quartz fragments. The
sections can be seen in prospecting pits.

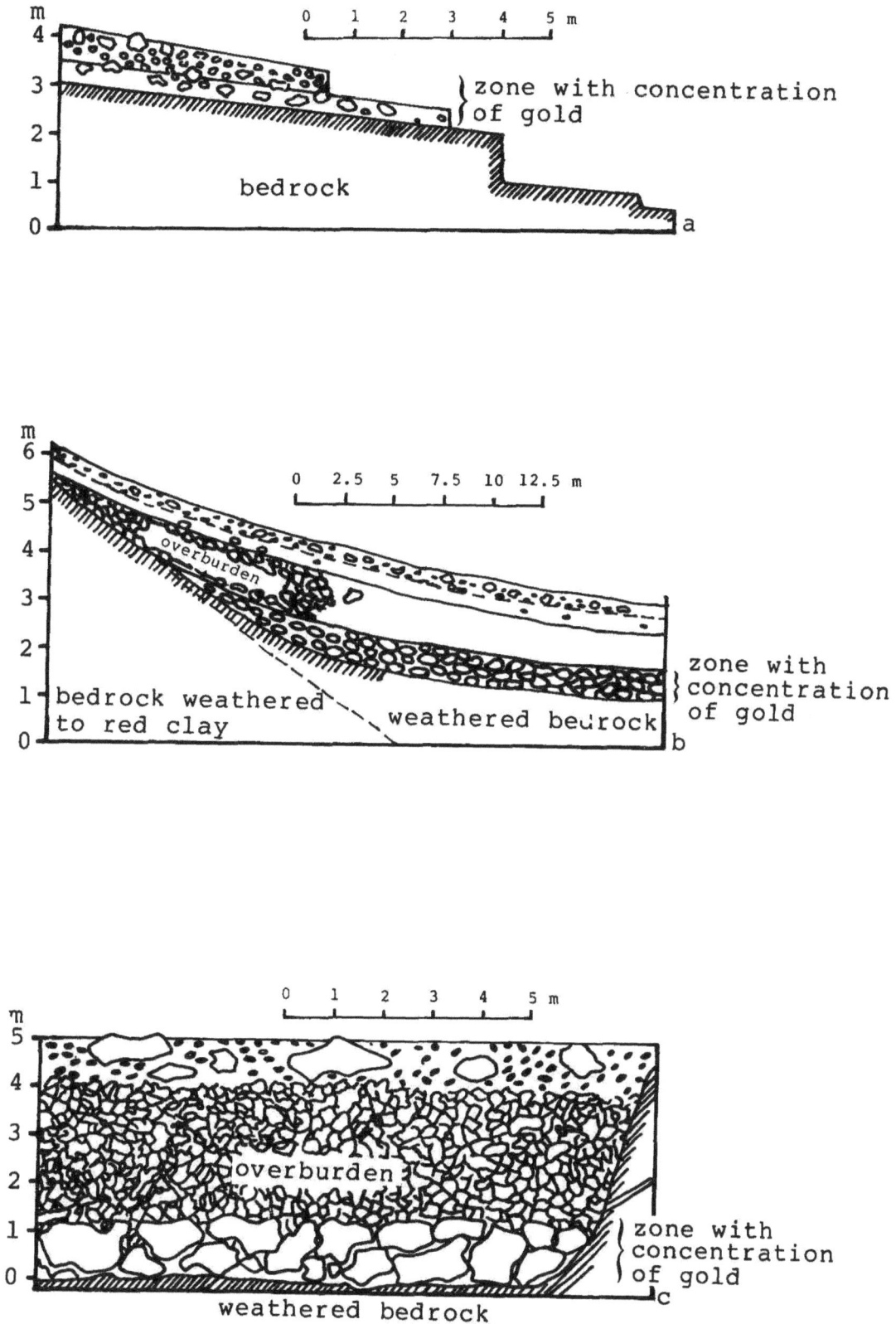

Fig. 5 Different types of gold bearing
alluvial deposits
(According to Brinck, 1955)

There are two types of alluvial deposits:-

- the older stream and river conglomerates which sometimes form terraces to depths of 80 metres above the stream bed, as shown in Fig. 6.

- the young gravel and sand deposits in recent streams and rivers as shown in Fig. 7.

The greatest gold concentrations are found at the base of the gravels and conglomerates and in the top 2cm of the underlying clay bedrock. If the bedrock is hard and not weathered then there can be gold enrichment in cracks, grooves, holes and other rough places of the river bottom.

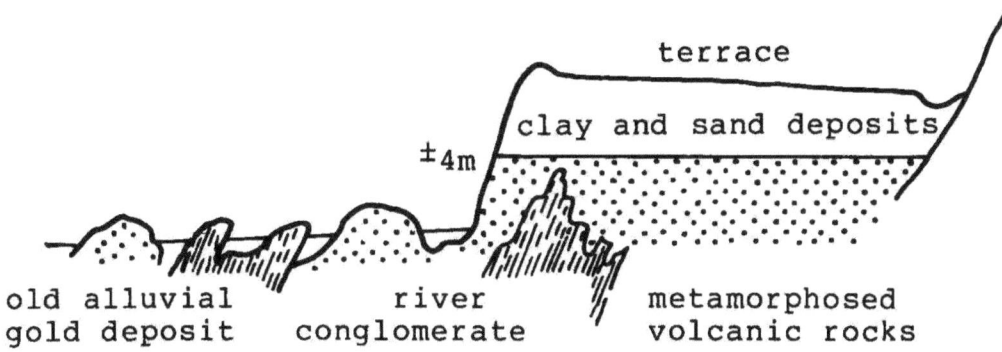

Fig. 6 River conglomerate forming a terrace
as part of an old gold deposit
(According to Brinck, 1955)

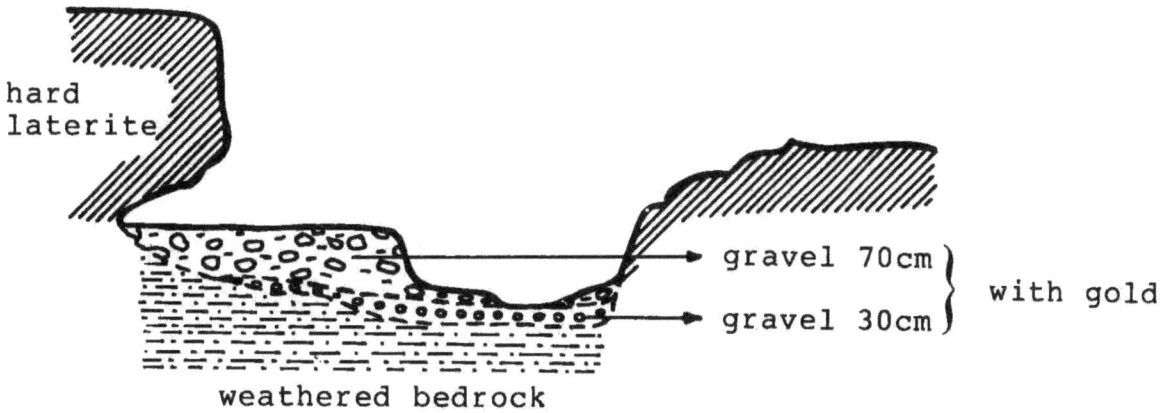

Fig. 7 Stream with young gold bearing deposit
(According to Brinck, 1955)

Fig. 8 Concentration of material with a batea

4.   EXPLORATION FOR GOLD

When searching for gold, there are two phases -
prospecting and exploration. Before talking
about the different prospecting and exploration
methods the techniques and equipment which are
used for the treatment of the ore and
concentration of gold will be discussed.

4.1   Gold Concentrating Methods

As part of the support facilities for people who
would like to undertake the mining of gold, the
GMD has established facilities at Loksi Hati on
the Saramacca river and at Jorka Creek on the
Maroijne river where experienced GMD personnel
give instructions in the various methods.

These consist of map reading, building of small
camps, simple administration, digging of
prospecting pits, de-sliming and various
concentration techniques. In addition,
instruction is given in the construction,
maintenance and operation of the various types of
equipment.

During the prospecting and exploration phase the
students can also obtain advice from experienced
prospectors and use some of the GMD facilities,
such as mail services. Nominal costs are
charged, depending on circumstances.

4.1.1   The Batea

The batea is a circular metal pan with a pointed
bottom, similar to a Chinese hat, as shown in
Fig. 8. Gold and heavy minerals are concentrated
in the bottom of the pan, which is floated in
slowly flowing water and is given a circular
eccentric motion so that the light minerals spill
over the edge into the river. Before starting
the process it is necessary to remove all
particles greater than 2.5cm in diameter and
organic matter. The contents of a batea with a
diameter of 45cm, such as used by the GMD, is 6
litres. An experienced man can treat $0.5m^3$ per
day. Because of this low throughput only very
rich deposits can be treated with a batea.

4.1.2   The Sluicebox

This piece of equipment has made the greatest
contribution to the gold industry in Suriname and
the rest of the world in the treatment of
secondary alluvial and colluvial deposits. Its
advantage is that it can handle greater
quantities than the batea, so that deposits with
lower gold contents can be treated. Another very
important advantage is that finer gold particles
can be recovered than with a batea. In a longtom
operation the separation is achieved by a stream
of water running in a launder in which slats are
fitted to form riffles, so that the gold and
heavy minerals are retained and the light
minerals such as quartz are washed away. The
construction is shown in Figs. 9, 10 and 11.

A sluicebox normally has a length of 2.5 to 3.5m
and width of 15 to 100 cm. It has been found
that two or three boxes in series will recover
90% of the mined gold (+1.5mm) of which 75% stays
in the first box. If the gold is very fine then
the recovery can be improved by placing more
boxes in series, or by increasing considerably
the width of the second box or by using an
undercurrent. In paragraph 4.1.3 a description
of the undercurrent is given. GMD uses a
threefold installation as shown in Fig. 9. In
the first part which is called the torpedo the
clay lumps are broken up with a flat wooden
implement which is moved from side to whilst
water is being added. A screen is fitted before
the first sluicebox which allows only particles
of less than 2.5cm to pass through. A 1cm screen
is fitted before the second sluicebox. To retain
the fine gold which passes the second sluicebox,
the material flows through a final gold trap
where a detergent such as liquid soap is added.
A jute sack or fine coconut mat is fitted to the
bottom of this box and this retains the fine
gold.

In a two-man operation in which the material is
shovelled into the longtom, care has to be taken
to have available sufficient water to treat 2.3m³
during a 10 hour working day. In this case
approximately 200 litres/minute of water (which
cannot be re-cycled) is required. If the water
cannot be fed by gravity then a 2 inch pump has
to be used.

-14-

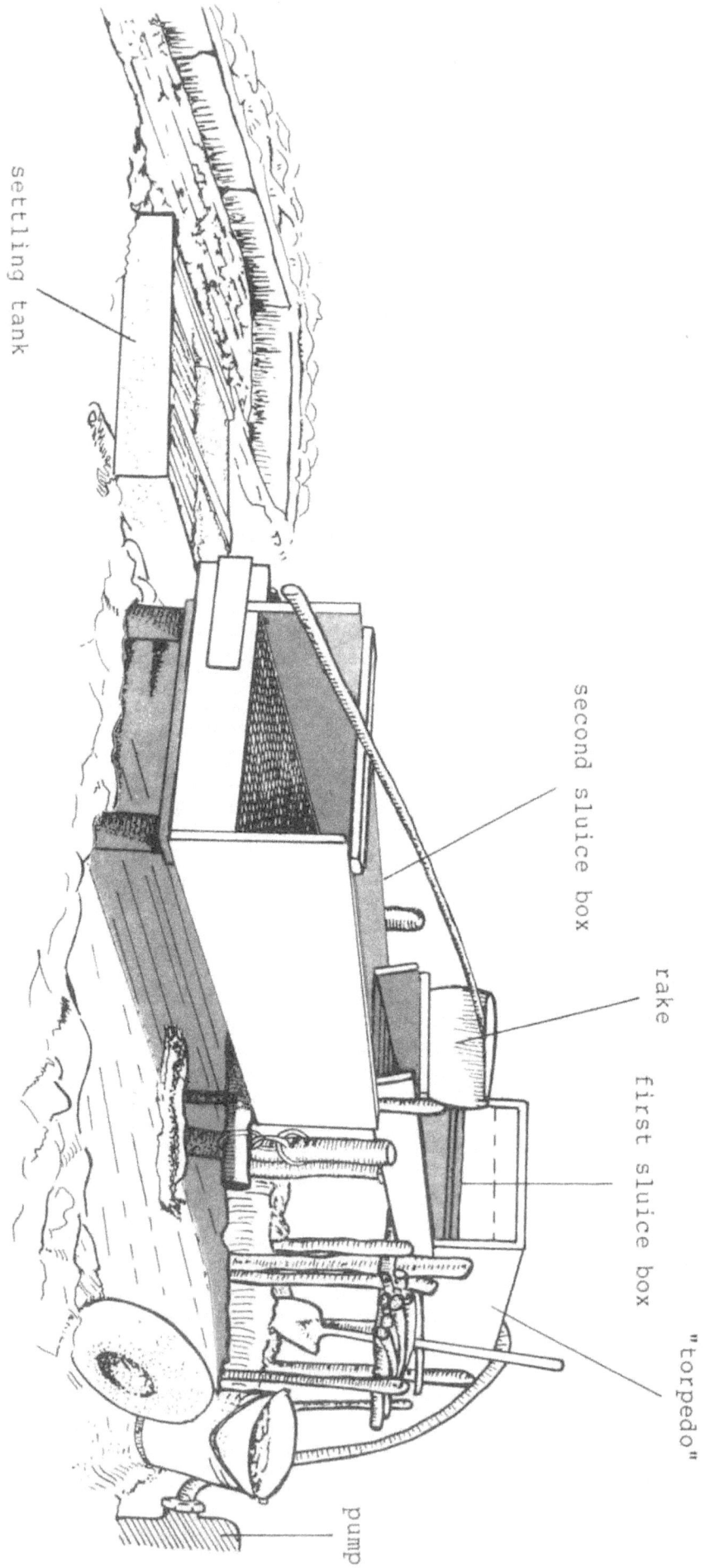

Fig. 9   Arrangement of sluicebox as used by GMD

settling tank

second sluice box

rake

first sluice box

"torpedo"

pump

175cm

175cm

40cm

56cm

70cm

40cm

1 off

25cm

35cm

5cm

1 off

64cm

3cm

It is recommended that the bottom be covered
with an old piece of conveyor belting

board thickness 2cm

175cm

175cm

25cm

2 off

2 off

150cm

30cm

66cm

150cm

80cm

1 off

25cm
2 off
10cm

1cm
aperture
mesh

35cm

1 off

80cm

2 off

25cm

30cm

150cm

150cm

4 off

25cm

1 off

84cm

76cm

76cm

1 off

25cm

1 off

10cm

84cm

2 off

10cm

150cm   **4 off**

5cm
5cm

notches 2cm deep

scale 1 : 200

5cm
5cm
10cm

5cm

10cm
5cm

10cm

15cm

12 off

scale 1:1
aluminium strips
10 off

2cm

86cm long

3cm

mats or white jute bags 76x150cm
80x150cm

water supply

de-sliming tank

screen

long tom or
riffles box

(not to scale)

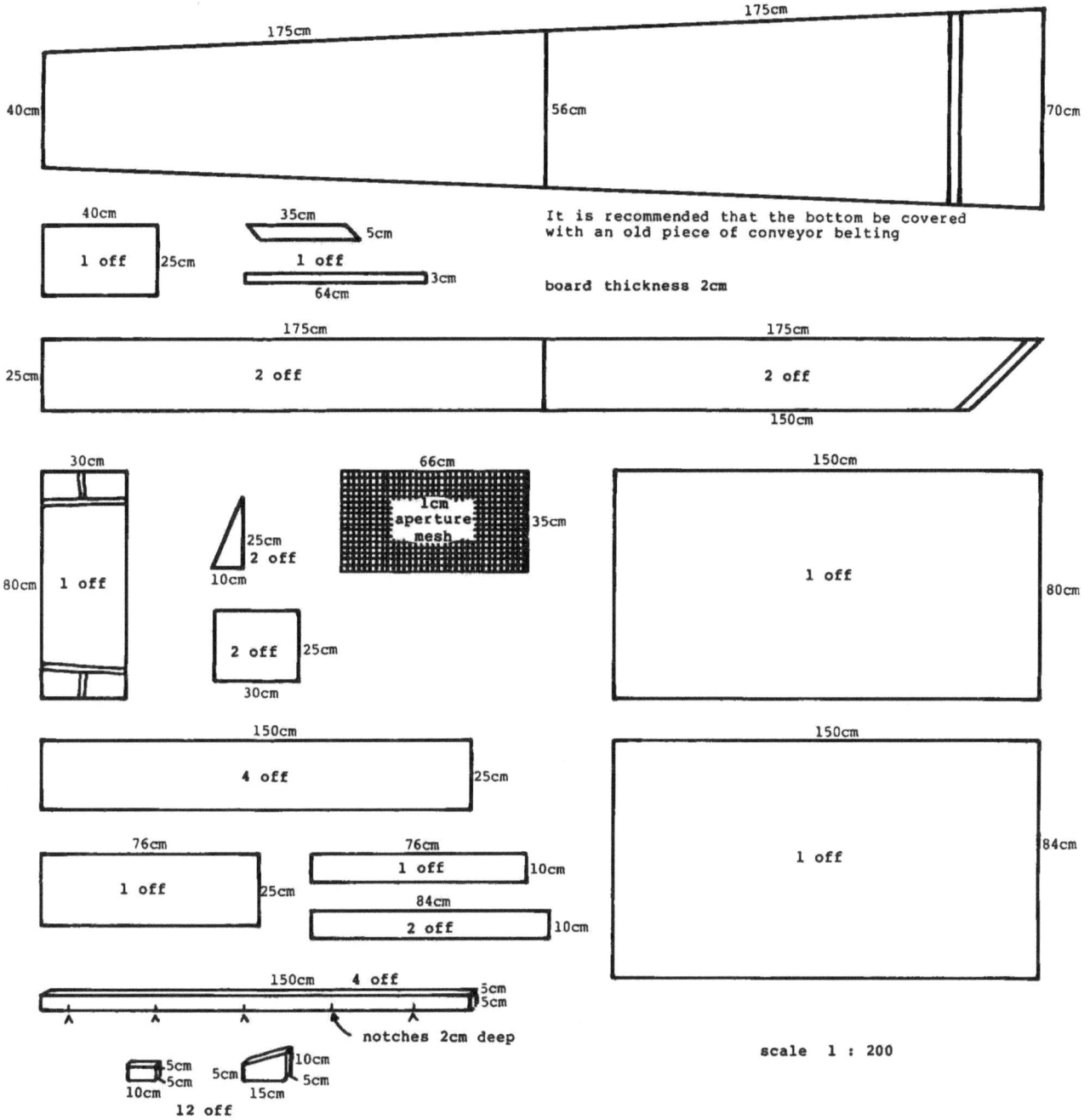

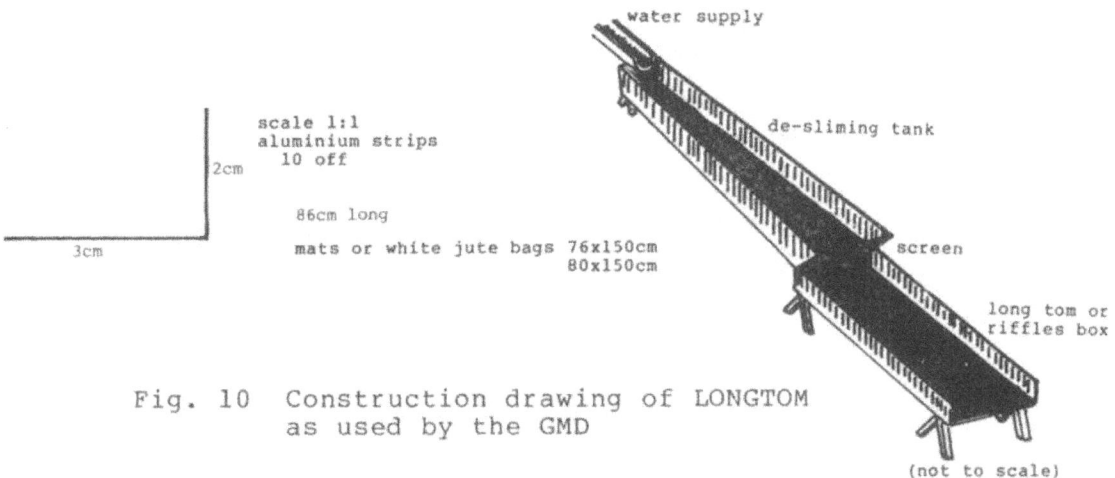

Fig. 10   Construction drawing of LONGTOM
as used by the GMD

-16-

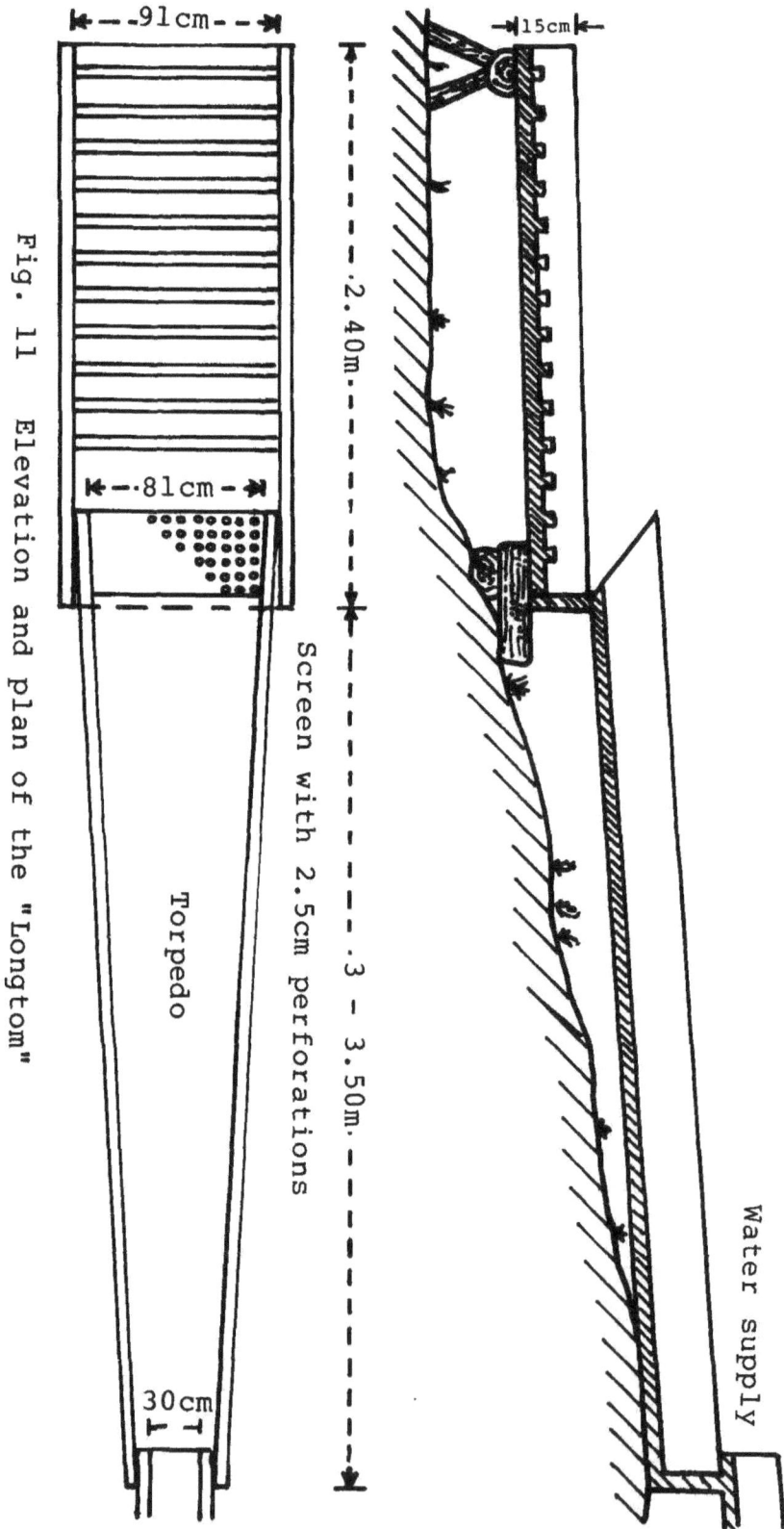

Fig. 11   Elevation and plan of the "Longtom"

The amount of water running in a stream can be estimated by throwing a piece of wood into it and measuring the distance moved in one minute. The average width and depth of the stream has to be measured and then the quantity of water flowing per minute equals 0.75 x depth x width x distance travelled per minute.

Example:

Average width of stream	=	1.5m
Average depth	=	0.75m
Cross sectional area	=	1.125m²

If the piece of wood travelled 10 metres in one minute then the quantity flowing is 0.75 x 1.125 x 10 = 8.44 m³/min (8440 l/min).

The slope of the sluicebox is very important and is usually about 4%, or 4cm/meter length. If the material contains a lot of clay, and the clay has to be broken up then the slope can be 6-8.5%.

It is very important to check that the clay has not formed small balls because gold can be caught up in these and will be washed out of the last sluice box.

If the riffles are filled with sand then the slope is too small, but if the gold is washed away then it is too big. The ideal slope will be found experimentally, and it is better to start with a slope which is too steep rather than with one which is not steep enough.

The heavy minerals containing the gold will have to be removed from the longtom at regular intervals. During prospecting and exploration this will have to take place after the treatment of each pit, but during exploitation the frequency depends on the heavy minerals content of the deposit. In the beginning it is important to clean up every two days, but later one should clean up as infrequently as possible, because it takes time and reduces productivity. When cleaning up, the tops of the riffles are first washed clean of gravel and sand, after which the flow of water is stopped. Then the riffles in the upper boxes have to be cleaned carefully with a hand duster to remove all the gold adhering to them. After that a stick of 2.5cm by 5cm is fitted across the outflow of the box and the water turned on just sufficiently to transport the gold and heavy minerals. This is assisted by a flat wooden spoon with a V-shaped edge which is moved up and down at an angle of 45° with the bottom of the box (which has to be very smooth) so that the gold and heavier minerals are concentrated at the top end of the box and the other heavy minerals are washed away. Any remaining light minerals are removed by hand. The gold and heavy minerals are put into buckets for further concentration in a rocker or batea or sent to a central plant in heavy lockable drums for mechanical concentration. Concentrates from the lower boxes are treated in the same way as those from the top box. If mats are used, they are washed in suitable containers, and the contents are further concentrated.

4.1.3   The Undercurrent Box

The working principle of the undercurrent box is that the material, after removal of the coarser fractions, is spread out over a wider shallower stream, the slope of which is steeper, but the water addition reduced so that the fine gold is retained on coir mats, animal pelts, or corduroy materials. Alternatively, riffles are placed at right angles to the direction of the flow. The undercurrent box is placed at right angles to the sluice boxes as shown in Fig. 12. Between the sluiceboxes and the undercurrent box is a grizzley, 1.2m long and as wide as the box. The bars of the grizzley are 2.5cm wide and the gaps 0.5 to 1.25cm. The fine material passes through the grizzley into a distributing box and from there into the two parallel undercurrent boxes having a total width of 2.5 to 3m. The slope of the undercurrent boxes is relatively steep - 8 to 12.5% - whilst the distributing box is even steeper.

4.1.4   Ground Sluicing   (Fig. 13)

This method is particularly suitable for the treatment of deposits which are 2 - 2.5m thick from which barren overburden has to be removed.

One condition is that the bedrock must have a slope of at least 4 degrees and that there is enough space for the tailings which remain after the treatment.

1000 to 2500 litres/minute of water is made to run in a stream through the deposit so that the overburden and the gravel are washed away, leaving behind the gold and heavy minerals in the irregularities of the bedrock.

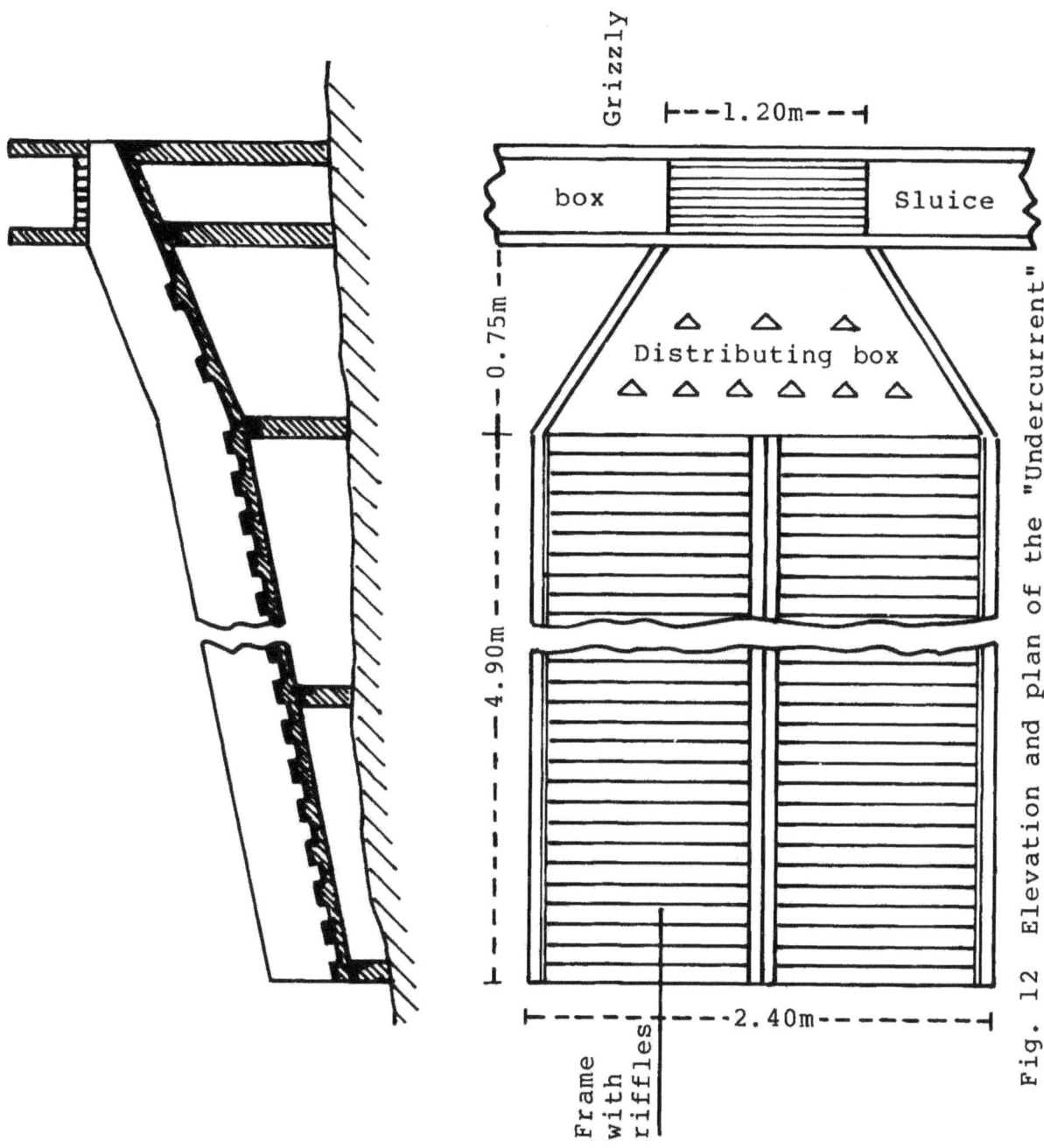

Fig. 12 Elevation and plan of the "Undercurrent"

cleaned surface
of bedrock

B

D

S

stream bank

M → sluice boxes

A-A = cross section
B = dump of large stones
D = earth dyke
M = sluice box
S = stream
T = trench

original surface

T

B

D

S

gravel

bedrock

Section A-A

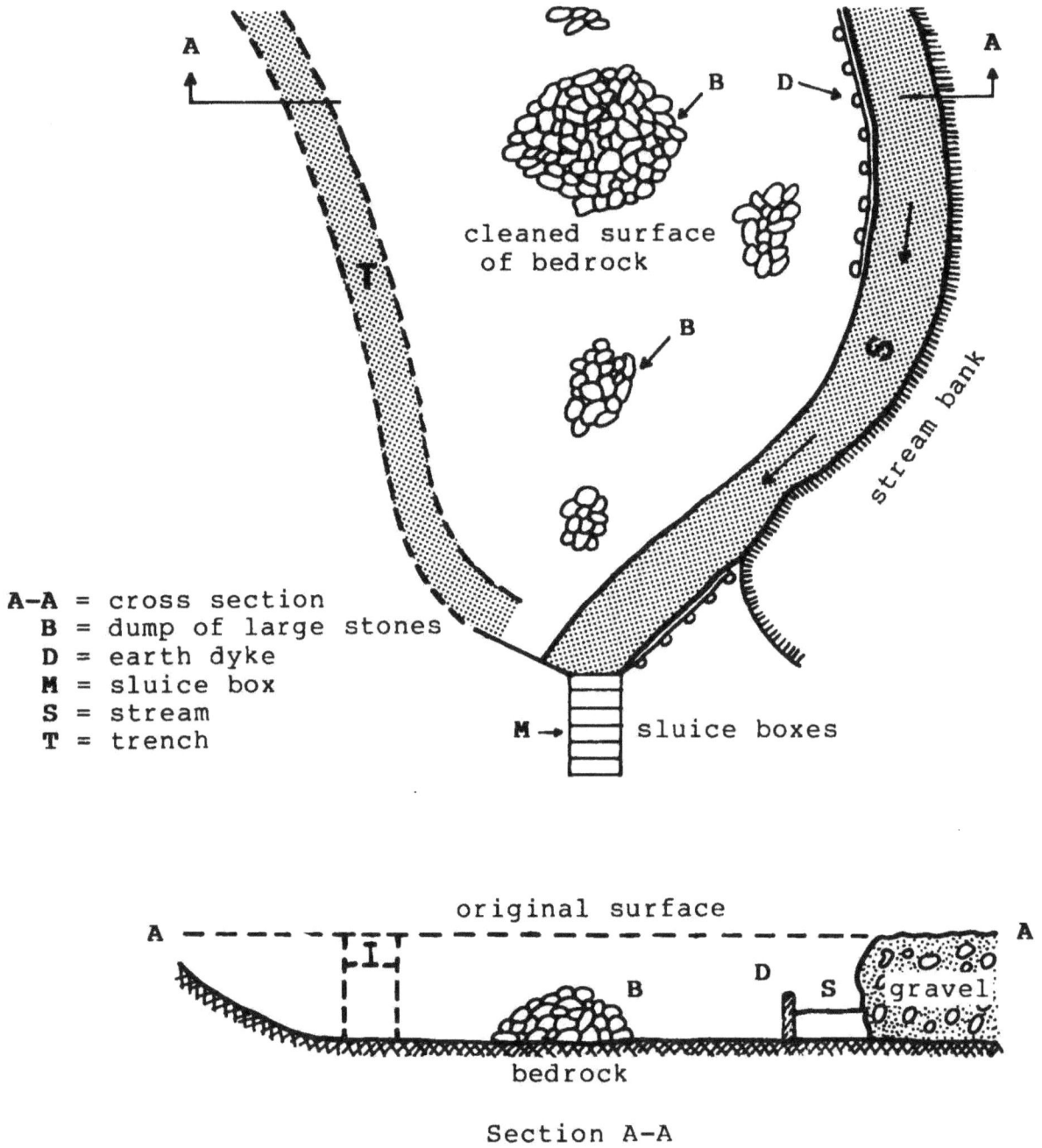

Fig. 13   Groundsluicing

4.1.5 The Rocker (Fig. 14)

This consists of a box in the form of a trough with riffles. The box is mounted on curved rocker bars. Above the box is a hopper with a screen on the underside to guide the material into the rocker. Under the rocker is a piece of cloth in which the fine gold is caught. The average size of a rocker is 1.37 to 1.52m wide by 0.61m long and 0.46m high. The wood of which it is made must be (as in the case of the sluicebox) a single piece free of knots and other irregularities, and must be completely flat. The difference in height between the front and back is 7.5cm. By swinging the rocker vigorously under a steady stream of water the gold is retained and the sand washed out. A rocker is best operated by two men who can handle 2 cubic meters in a 10 hour day. About 600 to 1100 litres of water are required, which can be recycled via a 200 litre container.

4.1.6 Suction Dredge (Fig. 15)

This is a modern dredging method whereby the gravel and gold are sucked up from the stream or river bottom, rather like a vacuum cleaner.

The transport medium is water. The installation consists of a raft, sluicebox, engine, pump and suction pipe with inlet. The diameter of the suction hose varies from 35 to 200mm diameter, throughput from 3 to 22.5 cubic metres/hour, the maximum dredging depth from 2 - 10m, the weight of the installation, 15 - 340 kg and the fuel consumption 0.83 to 0.34 litres per cubic meter. If the depth of water exceeds 1m the suction inlet is operated by a diver who can stay under water for several hours. In such cases, a compressor must also be provided. The heavy minerals containing the gold are retained by the riffles, periodically removed from the sluicebox and further concentrated in a batea, superpanner or jig. Because of the relatively high velocity with which the material passes through the sluicebox, care has to be taken when working deposits with a high content of fine gold not to lose some of the gold back into the river. It is possible to consider undercurrent installations and gold traps combined with heapleaching methods.

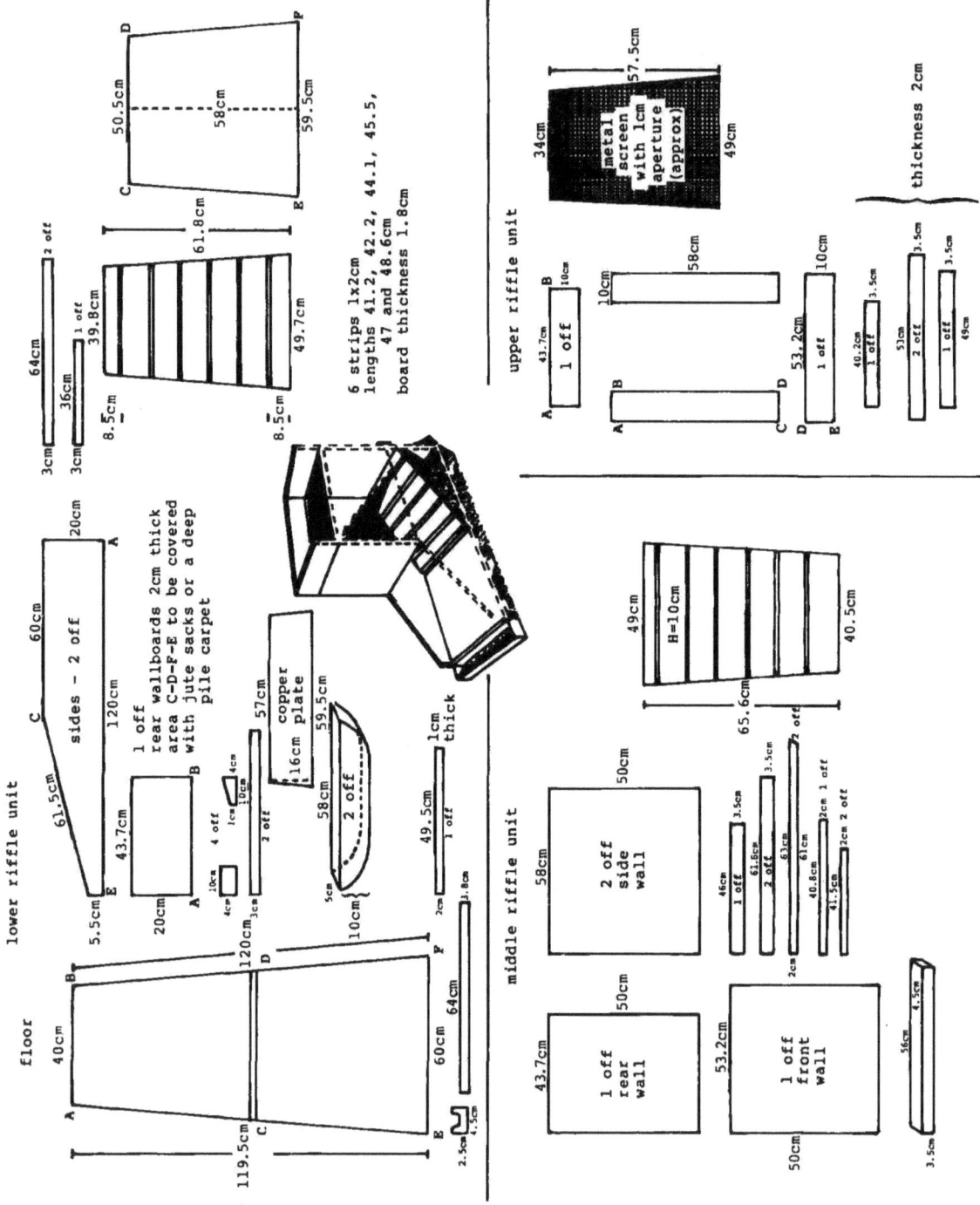

Fig. 14   Construction drawing of ROCKER in use by GMD

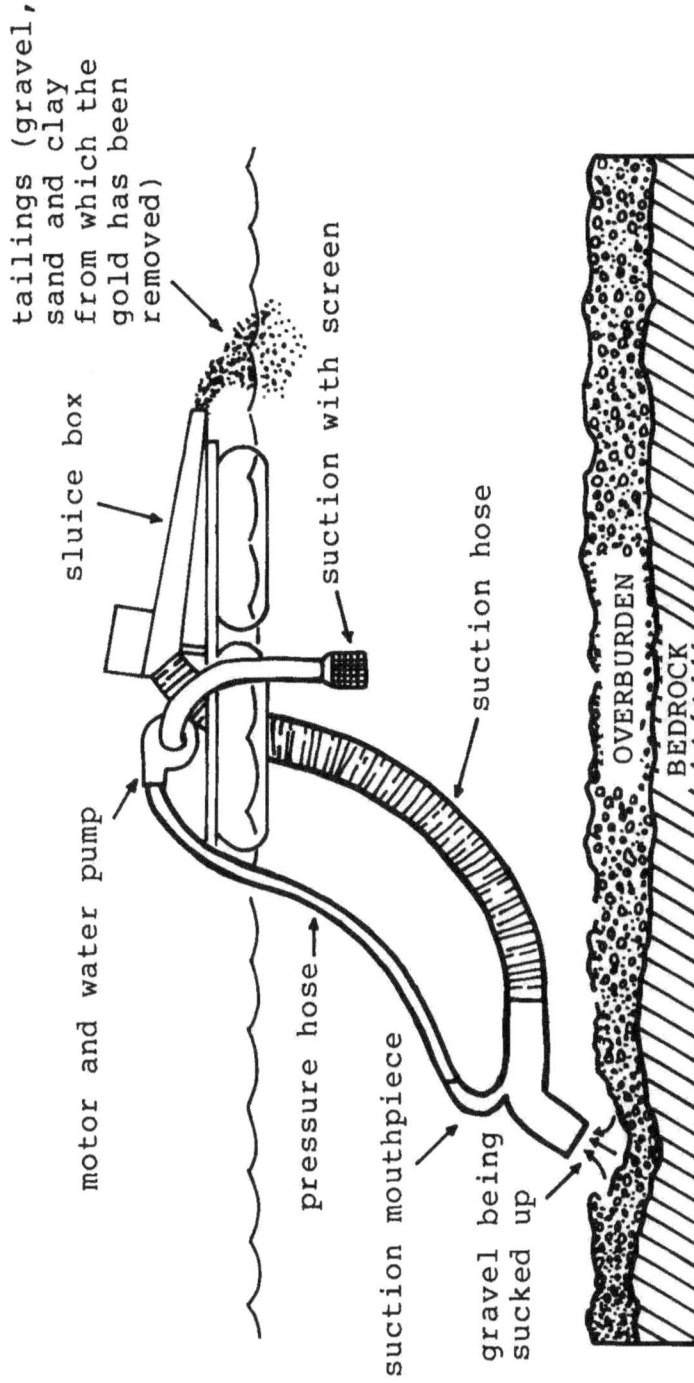

Fig. 15 Suction dredge
(According to Thornton, 1977)

It can be seen that a suction dredge can treat
relatively large quantities, so that it is able
to work a deposit which would be marginal when
using a longtom. The smallest type of suction
dredge can treat 30m^3/day compared with 2 - 3m^3
/day when two men are feeding a longtom by
shovel.

In all cases, be it sluicebox, ground sluicing,
rocker or suction dredge, it is necessary to
separate the gold from the heavy minerals which
have been retained by the riffles. In small
scale mining the simplest method of final
concentration is by a batea worked in a 50 litre
oil drum cut in half and filled with water, so
that the light minerals, if required, can be
re-concentrated several times until they no
longer contain any gold.

Another method is to store all the heavy mineral
concentrate and to send it periodically to a
central plant which is run on a co-operative
basis and where the gold is separated by
mechanical methods such as a superpanner or jig.

It pays to take periodic samples of the heavy
mineral concentrate from which the gold has been
removed and send them to the GMD laboratory,
because there is a real possibility that they may
contain valuable minerals such as tin, platinum,
diamonds and other precious stones.

If the gold is very fine and is lost in the
concentration process in the batea, longtom or
rocker then the heavy minerals can be treated
with mercury. Gold and mercury combine in a
ratio of 1:1 to form an amalgam. Mercury is
separated from the gold by evaporation as
explained in Chapter 6.

4.1.7 Leaching

(a) Heapleaching (Fig. 16)

This is fundamentally different from
the previously described gravitational
methods. In heapleaching, the gold
bearing material, which has to have
good permeability, is placed in large
heaps on an impermeable base. It is
sprayed with a sodium cyanide solution
which is recycled for several weeks or
months until all the gold is

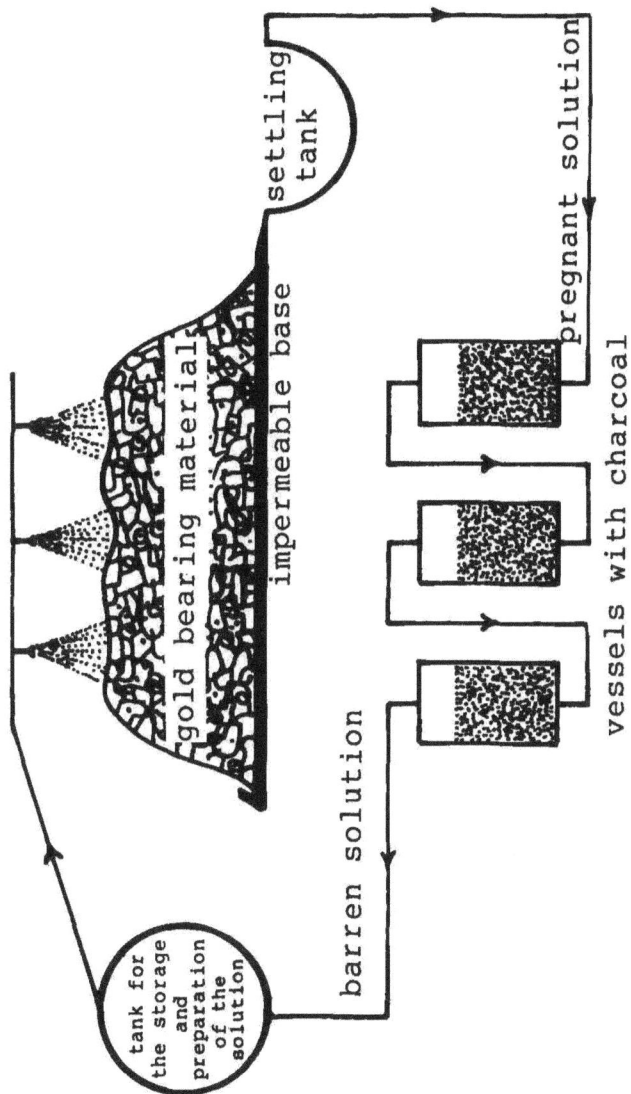

Fig. 16 Heapleaching arrangement

dissolved. Progress is checked by testing the gold content in the circulating cyanide solution. When the gold content no longer rises then it can be assumed that all the gold has gone into solution. The so-called 'pregnant solution' is passed through retorts containing charcoal or activated carbon which absorbs the gold. The charcoal is then burnt and the gold melted out. This method uses very poisonous solutions and is to be introduced into Suriname in the near future by experts who will operate it during the early periods. It will be obvious that this method is only suitable for larger operations, perhaps managed by co-operatives.

(b) Agitation Leaching

The method will recover even the finest gold. Gold-bearing clay can also be treated provided it is slurried and kept in suspension in large tanks.

(c) Vat Leaching

The vat leaching method, on the other hand, is more suitable for recovering gold from sands which have not been ground fine enough to be recovered by the use of agitators. Vat leaching uses a tank with duck boards in the bottom covered with hessian cloth. This in turn is covered with a layer of finely woven coir matting. When ready the vat is filled with ore and cyanide solution is then fed into the tank until all the ore is covered with solution. Samples are taken, and when the gold content shows acceptable levels of dissolution the discharge valves are opened and the pregnant solution is sent to the precipitation unit. Samples are taken at regular intervals until the gold content in the solution shows that an acceptable recovery has been achieved. The vat is then emptied and the cycle re-commenced.

4.2   Prospecting

If it is suspected that gold occurs in an area
then this can be verified relatively quickly by
prospecting. At this stage of the investigation,
there is no need to determine the actual amount
of gold in the ground, but it is necessary to
differentiate between meaningful indications and
those which need further investigation. We shall
return to this subject after discussing different
methods of prospecting. There are two such
methods - geochemical and gravity.

4.2.1   Geochemical Prospecting

In geochemical prospecting samples are collected
systematically on a grid basis and sent to the
GMD laboratories for gold determination. The
samples are taken at fixed intervals along grid
lines at a depth of 30 cm below the top soil, as
shown in Fig. 17. A grid of 400m x 20m is
recommended.

The samples of about 250g are packed in plastic
bags, 40 by 20cm and tagged with a label which
must contain the following information:-

   -    name and registration number of
        concession holder.

   -    name of the concession.

   -    line and picket number of sample
        location.

The samples must be accompanied by a sample list
and a map of the locations.

If 10 or 20% of a series of samples contain gold
of more than 0.2 ppm or 0.2g/tonne then the area
has an economic potential. If the percentage is
less but the area appears to extend beyond the
boundaries of the prospected area, as shown in
Fig. 17, then it is worth extending the area
under investigation in the indicated
direction(s).

4.2.2   Prospecting using Gravity Concentration Methods

In this method of prospecting, gravel and sand
samples are collected every 250 metres, as shown
in Fig. 17, and are concentrated in a batea.

Line A   B   C   D   E   F   G   H

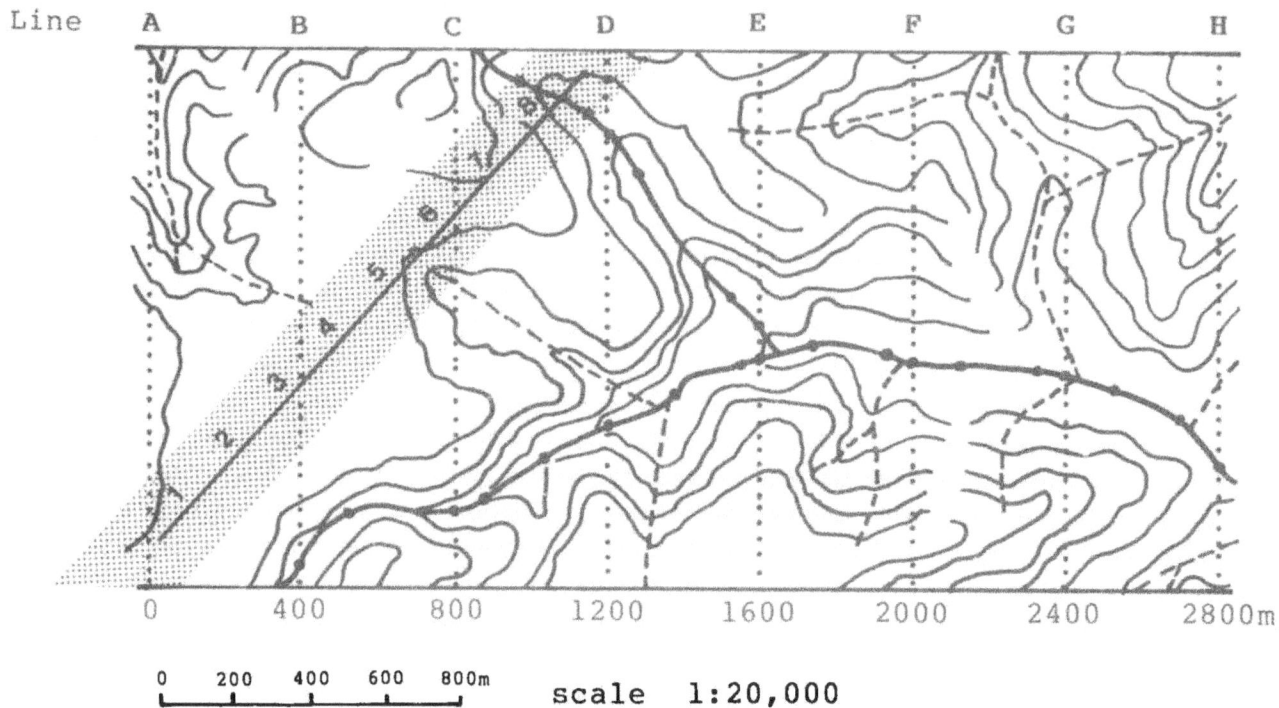

scale 1:20,000

▲   location of gold bearing quartz dyke

▒   zone with soil samples containing more than 0.2 ppm of gold

●   location of batea samples

    location of rock chip samples

    Fig. 17

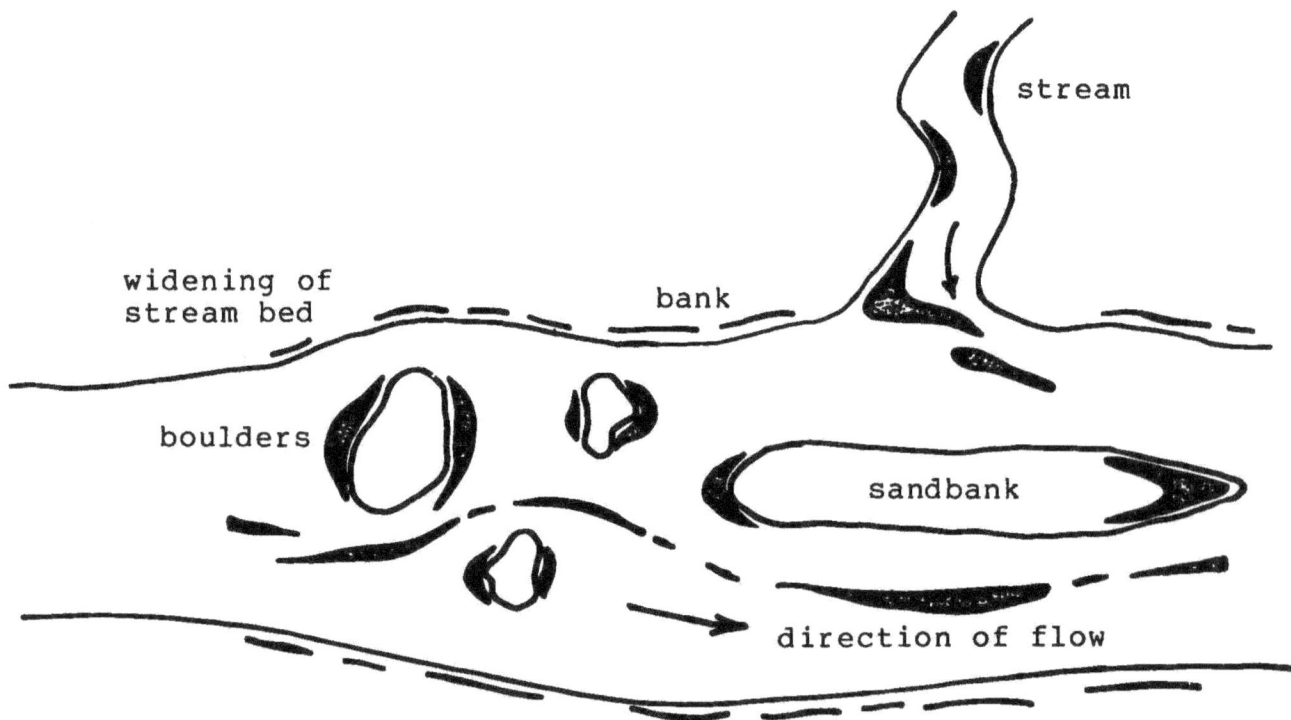

Fig. 18   Locations where concentrations of gold and
          heavy minerals can be expected. Concentration
          is greatest underneath boulders and other
          obstacles of the widening stream bed.
          (According to Zeschke, 1964)

Before starting with this labour intensive way of
collecting information, it is worth taking a few
samples at the most likely locations, as shown in
Fig.18. If no gold shows in the batea it is then
necessary to dig pits down to bedrock at the
places shown in Fig. 18 and to treat samples of
the sand and gravel immediately above bedrock and
from the top layer of the bedrock itself.

In order to obtain an indication of the gold
content one can use reference pictures supplied
by GMD which consist of plastic cards showing the
particle sizes and average weights encountered.
It is possible to overestimate the amount of gold
by a factor of two or three because it frequently
occurs in platelets.

If one starts from the fact that a batea of 45cm
diameter contains 6 litres or 1/167th part of a
cubic metre of gravel then it is possible to
determine the gold content per cubic metre, as
the example in the following table shows:

Class	Size mm	Average Weight mg.	Proportion	Weight of Fraction mg.
1	+2	80.8	0	0
2	2 – 0.85	8.61	0	0
3	0.85 – 0.425	1.84	2	3.68
4	0.425 – 0.25	0.42	3	1.26
5	– 0.25	0.09	17	1.53
		Total		6.47 mg

The gold content of this material will therefore
be 167 x 6.47 = 1080 mg./m^3 or 1.08 g/m^3.
Records of results should be kept in a very clear
manner, as illustrated in Fig. 19, and all the
gold recovered in pits or streams must be handed
over to the Head of the GMD.

If 10 - 20% of the pits or surface samples
contain gold, as averaged in Fig. 19, then the
area could be interesting and further exploration
would be justified.

It is important in geochemical prospecting that a
sample should be treated in a batea at each point
where a stream or river crosses a sample line.

If the prospecting phase has been successful and
an area has been designated as positive then it
is necessary to determine the type of deposit -

BATEA PROSPECTING

Name of concession holder . . . . . .
Area of concession . . . . . .
Concession No. . . . . .
Name of batea operator . . . . . .

Pit No. . . . .

Sample No.	Depth in cm	Clay	Sand	Gravel	Occurrence of gold particles by class*					Tot.wt of gold mg	x163= mg/m³	Vol.% gravel +7-10 cm	mg gold per m³ after correction	Calc. gold per layer	Remarks
					+2 mm 6.00mg	2 mm 0.85mm 1.56mg	0.85mm 0.425mm 0.31mg	0.425mm 0.25mm	-0.25mm 0.05mg						
1	0–20	x			–	–	–	–	4	0.2	33.4	0	33.4	13.36	Groundwater at a depth of 85cm
2	20–80	x̲	x		–	–	2	5	10	0.82	136.94	0	136.94	109.55	x̲ = mainly
3	80–130	x	x		–	–	5	5	10	2.05	342.35	0	342.35	342.35	
4	130–180	x̲	x	x̲	(10mg)	5	3	3	10	46.11	7700.37	15%	7700.37– 1155.06= 6545.31	7700.37– 1155.06= 6545.31	
5	180–200	x		–	–	–	–	–	–	–	–	–	0.0	0.0	

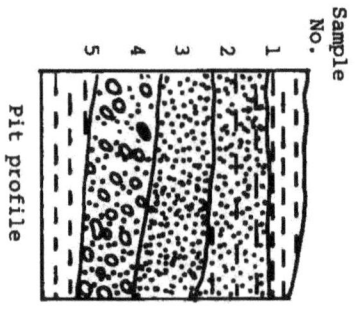

Total 7010.57mg

Average gold content in pit 7010.57 ÷ 5 = 1402.11mg/m³ or 1.4g/m³
Thickness of overburden 1.30m
Thickness of gold bearing layer 0.50m

* The average weight of gold particles varies from area to area depending on their shape

2* The volume of material in sample no. 4 in which 7,700.37mg of gold were found was 15% greater and therefore the weight of gold 15% larger. 0.15 x 7,700.37 = 1.556.06mg. The true content was 7,700.37 – 1,556.06 = 6545.31mg/m³

3* The content was recalculated for the thickest layer, sample no. 3, of 50cm. If the gold in sample no. 1 of a thickness of 20cm is spread over a thickness of 50cm, then the grade is reduced in the proportion of 20/50=0.4; the grade will be 33.4 x 0.4 = 13.36mg/m³

Sample No.
1
2
3
4
5

Pit profile

Fig. 19

whether it is a primary (a quartz reef) or a
secondary colluvial or a secondary stream
deposit-as this will affect the work programme
for the next phase.

All quartz pieces which are found along the
sample lines or in their immediate vicinity
should be carefully recorded on a map, together
with a note of whether they were sharp or
rounded. It is also necessary to occasionally
break up large stones with a hammer, to grind
them in a mortar, and to determine in a batea
whether they contain gold. If gold-containing
quartz samples can be found along several lines,
as shown in Fig. 17, then the direction of the
quartz reef can be determined.

By observing the profiles of the pits and whether
the fragments of the gravel or conglomerates are
of approximately the same size and rounded or
with great size variations and sharp edges, the
nature of the deposit can be determined, i.e.
whether it is alluvial or colluvial.

4.3   Exploration

The objective of this phase is to obtain reliable
information on:

    -    the surface area of the gold deposit
         and its depth

    -    the gold content of the material to be
         mined

    -    the ways in which the gold particles
         occur in the deposit i.e.

         *   their size distribution

         *   the point of maximum gold
              concentration

         *   the nature of the gangue in which
              the gold occurs, whether clay,
              sand, gravel, quartz or a mixture
              of them.

The information obtained during the prospecting
phase will help us determine the exploration
phase according to the way the gold is occurring,
either as a primary or a secondary colluvial or
alluvial deposit. The basic principle during the
exploration phase is that the concentrating

method should be the same as during the
subsequent exploitation phase, so that the amount
of gold recovered is not less then expected.

4.3.1 Exploration of Secondary Deposits

4.3.1.1. Stream and Alluvial Deposits

Location of the Pits

The place where the maximum gold was found during
prospecting should be taken as the starting point
for exploration. From here a straight line
should be drawn along the axis of the valley, and
the occurrence and depth of gravel should be
established with a steel test rod every 5 - 10
meters.

If it is shown that gravel occurs across the
width of the valley as shown in Fig. 20, it is
recommended that further pits be dug on a regular
pattern, i.e. a line will be laid out either side
of the original line at a standard distance of
50m, which will then be sampled and tested by the
GMD to establish whether gravel occurs in those
locations. The pits are dug on a grid of 50 or
25m. Scratch samples are taken vertically down
the whole depth of the pit and treated in order
to determine how the gold content varies with
depth. Every 15cm a sample, 20cm x 20cm (which
approximately equals the contents of a batea), is
taken out of the side of the pit. The gold
content per cubic meter of the collected scratch
samples is estimated as shown in the section on
prospecting.

When digging a pit, the layer with the highest
gold content is shovelled on to a heap and
treated in a longtom. The gold obtained is
weighed and is expressed in terms of the volume
of the whole pit, as shown in Fig. 22. The
following information is immediately available
from this table:-

-    the thickness of the gold bearing
     layer

-    the gold content of this layer in g/m^3

-    the thickness of the overburden

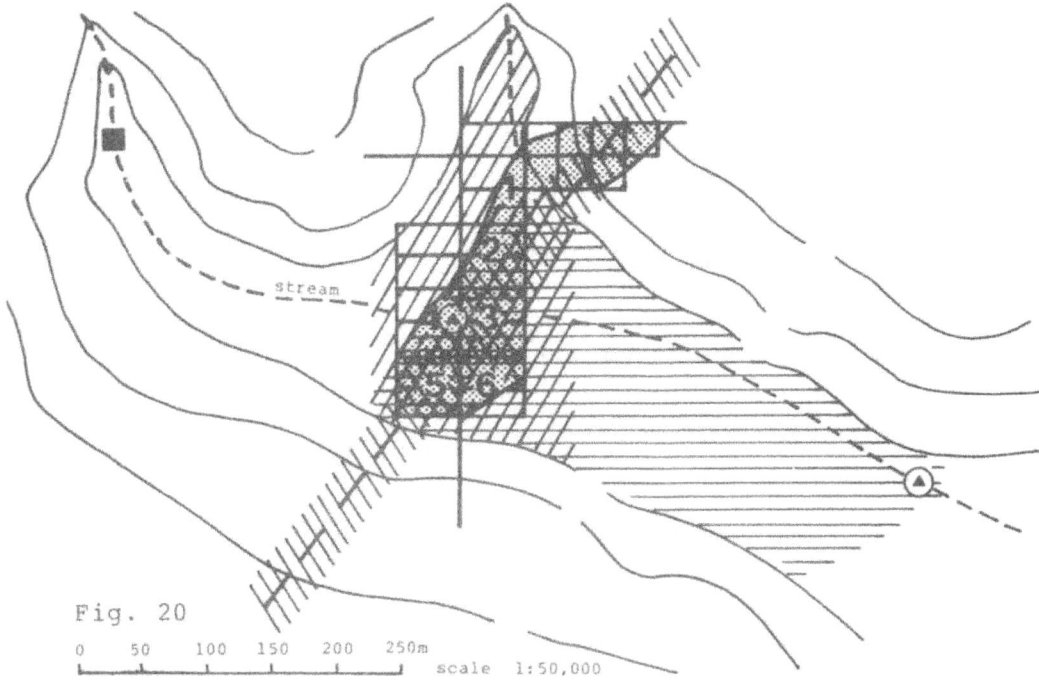

Fig. 20

scale 1:50,000

■ Location of prosp. pit Ⓐ Location of gold bearing prosp. pit

▨ Area where a steel probe indicated gravel

▩ Area where exploration of colluvial deposit should be considered

▤ Area where exploration of stream deposit should be considered

Area of possible exploitation

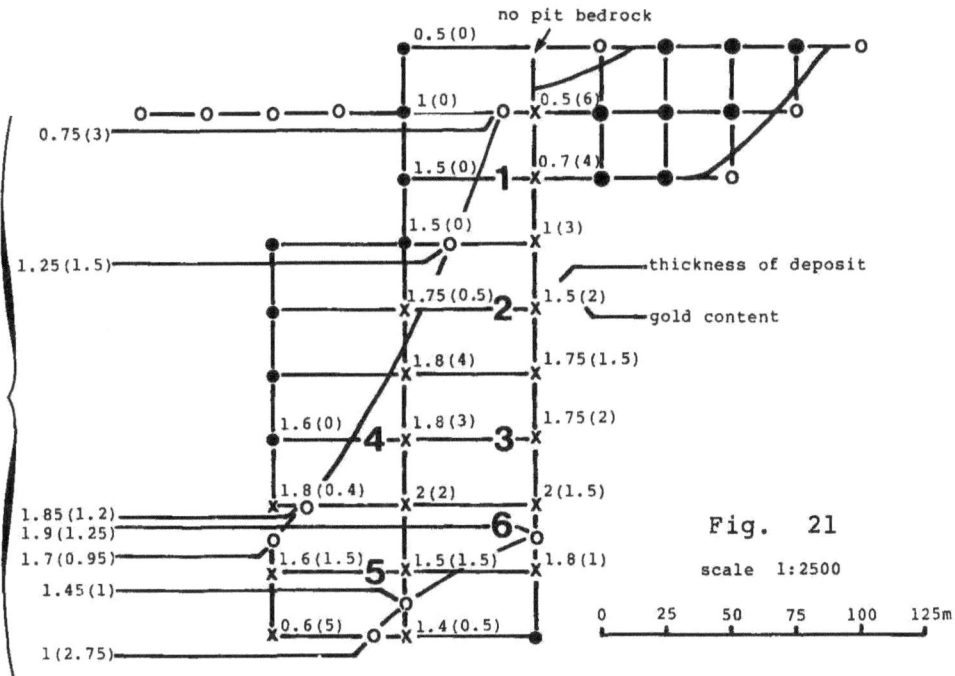

Fig. 21

scale 1:2500

thickness of deposit

gold content

X Location of exploration pits containing gold

● Barren pits

O Pits in colluvial deposits

● Ditto, containing gold

Primary quartz reef deposit containing gold

Name of concession holder . . . .   LONGTOM EXPLORATION
Concession area   . . . .
Concession No.   . . . .

PIT No. . . . .
Measurements 2x1x2m = 4m
Content 4000 litres

Sample No.	Content litres	Depth in cm	Clay	Sand	Gravel	Tot. wt of gold mg	Remarks
1	400	0-20	X				
2	1200	20-80	X	X			X = mainly
3	1050	80-130		X			
4	1000	130-180		X	X	8837.78	This means 8.84g/m^3
5	400	180-200	X				

Total   4000 litres   Total   8837.78

Average gold content in the pit $\frac{8837.78}{4000}$ = 2209.45mg/m^3 = 2.21g/m^3

Thickness of overburden   1.30m
Thickness of gold bearing layer   0.50m
Gold content of gold bearing layer   8.84g/m^3

Sample
No.

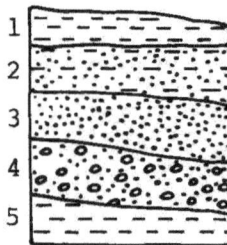

Pit profile   Fig. 22

- the average gold content in g/m³ of
the gold bearing layer plus the
overburden, i.e. of the whole pit.

If the gold content of the pits along the middle
line exceeds about 2g/m³ then exploration should
be continued by digging pits along the two
adjacent lines, and determining the gold content.
The location of the pits is recorded on a map
giving the distance between them as accurately as
possible. The thickness of the deposit and its
gold content is indicated at each location as
shown on Fig. 21.

Estimation of the Value of the Deposit

First, the limits of the deposit should be shown
by drawing a border inside which all pits have a
gold content of more than say 1.5g/m³ . The
border should consist as far as possible, of long
straight lines. To prevent an overestimation of
the value of the deposit it is recommended that
additional pits be dug in areas of high gold
content. This is also shown in Fig. 21. As
shown, the surface area is divided into blocks of
50m by 25m i.e. an area of 1250m each. These
are combined into double blocks. The area of
incomplete border blocks is simplified into
triangles and rectangles for estimation. The
surface area (a), the average thickness (d), the
volume (v), the average grade (g), and the value
(w) of each block are calculated and listed as in
Fig. 23. The total volume (V), the average grade
(G), and the total value (W) are shown at the
bottom of the table, Fig. 23.

The advantage of this method of calculation is
that, independently of the financial situation of
the project, it is possible to make a mining
programme giving the equipment to be used, the
rate of mining and a schedule of the sequence in
which the blocks are to be mined.

The minimum viable gold contents for treatment by
batea or longtom are 5.25 and 3.15 g/m³ respec-
tively. A suction dredge can only be considered
if enough water can be made available, say by
damming a stream.

The deposit shown (of 17,191 m³) can be mined
with a 2.5 inch suction dredge of 30m³ /day
capacity in 573 working days. Assuming that
there are 300 working days in a year then this
deposit will last for 1 year 11 months.

CALCULATION OF THE VALUE OF A DEPOSIT
(see also Figs. 21 and 22)

Block No.	O (= surface) (= number of squares x 1,250m²)	d (= average depth)	v (= volume)	g (= average content)	w (= v x g x g); p = f 12.50/g
1	1x1,250=1,250m²	0.50 m 1.00 m 1.25 m 0.75 m 3.50 m:4=0.88 m	1,250 m² x 0.88 m = 1,094 m³	0.5x6.00 = 3.00 1 x3.00 = 3.00 1.25x1.50 = 1.88 0.75x3.00 = 2.25 3.50 10.13:3.5= 2.89 g/m³	v x g x p = 1,094 x 2.89 x f 12.50 = f 39,520.75
2	1.75 x 1,250m²= 2,188 m²	1.00 m 1.75 m 1.80 m 1.25 m 5.80 m:4=1.45 m	2,188 m² x 1.45 m = 3,173 m³	1 x 3.0 = 3.00 1.75x1.5 = 2.63 1.8 x 4 = 7.20 1.25x1.5 = 1.88 5.8 14.71:5.8= 2.54 g/m³	v x g x p = 3,173 x 2.54 x f 12.50 = f 100,742.75
3	2 x 1,250m²= 2,500 m²	1.75 m 2.00 m 2.00 m 1.80 m 7.55 m:4=1.89 m	2,500 m² x 1.89 m = 4,719 m³	1.75x1.5 = 2.63 2 x 1.5 = 3.00 2 x 2 = 4.00 1.8 x4 = 7.20 7.55 16.83:7.55= 2.23 g/m³	v x g x p = 4,719 x 2.23 x f 12.50 = f 131,542.12
4	1 x 1,250m²= 1,250 m²	1.80 m 2.00 m 1.80 m 5.60 m:3=1.87 m	1,250 m² x 1.87 m = 2,333 m³	1.80x4 = 7.20 2 x 2 = 4.00 1.8x0.4 = 0.72 5.6 11.92:2.56= 2.13 g/m³	v x g x p = 2,333 x 2.13 x f 12.50 = f 62,116.13
5	2 x 1,250m²= 2,500 m²	2.00 m 1.45 m 1.00 m 0.60 m 1.70 m 1.85 m 8.60 m:6=1.43 m	2,500 m² x 1.43 m = 3,575 m³	2 x 2 = 4.00 1.45x1 = 1.45 1 x 2.75 = 2.75 0.6 x 5 = 3.00 1.7 x 0.95 = 1.62 1.85x1.2 = 2.22 8.60 15.04:8.6 = 1.75 g/m³	v x g x p = 3,575 x 1.75 x f 12.50 = f 78,203.13
6	1 x 1,250m²= 1,250 m²	2.00 m 1.90 m 1.45 m 2.00 m 7.35 m:4=1.84 m	1,250 m² x 1.84 m = 2,297 m³	2 x 1.5 = 3.00 1.9 x1.25 = 2.38 1.45x 1 = 1.45 2 x 2 = 4.00 7.35 10.83:7.35= 1.49 g/m³	v x g x p = 2,297 x 1.47 x f 12.50 = f 42,207.38

Total volume V = 17,191 m³   Average content $G = \dfrac{W}{V \times 12.5}$ Total value   W = f 454,332.26

Note: The calculated value of the deposit has to be corrected for the purity of the gold. This correction factor (C) is about 0.9 in Suriname.

Fig. 23

Before considering the mining of this deposit
with a suction dredge it will be necessary to
establish experimentally how many grams of gold
per cubic meter can be recovered with it. The
exploitable value of the deposit will then be:

$$W' = V \times G' \times C \times P$$

Where G' is the grade as mined by the dredge and
C a correction factor for the purity of the gold
- which can be determined in the laboratory of
the GMD.

4.3.1.2. River Deposits

If it is found during prospecting that the sand
and gravel of the stream bed contains gold then
the chances are that the adjoining terraces will
also carry gold. This assumption must be
verified at the places shown in Figs. 18 and 24
when the water level is low, by digging pits into
the sand and gravel banks and testing the
materials as in paragraph 4.3.1.1. The roots of
the plants along the banks and islands should
also be washed and the earth and gravel
concentrated. The contents of potholes should
also be treated. If the surface of the bedrock
is exposed and dry, then several centimeters of
its depth should be removed, and all the sand and
gravel, whether it be cemented with iron oxide or
not, should be tested for gold. Fig. 25 shows
how cracks and holes can collect the gold.

If these tests are positive then the river bed
can be explored with a suction dredge (the type
used will depend on the depth of water and
thickness of the sand and gravel bed) by dredging
every 100m at right angles to the direction of
the flow.

The volume of sand and gravel sucked up per hour
must be determined (this will present some
practical difficulties) as well as the weight of
the gold recovered. The grade can then be
calculated by dividing the weight of gold by the
volume of sand.

When dredging it is very important that any
potholes, cracks and depressions in the bottom of
the river are emptied completely, cracks should
be broken open with a chisel, cleaned out and
gone over again with the dredge. Cracks and
splits in the banks which contain gold and
continue into the river need special attention.

bank

←direction←   ←of flow←

bedrock
outcrop

concentration of heavy
minerals and possibly
gold

bank

boulders and
other obstacles →

←direction ←   ←of flow←

bedrock
outcrop

bank

←direction ←   ←of flow ←

bedrock
outcrop

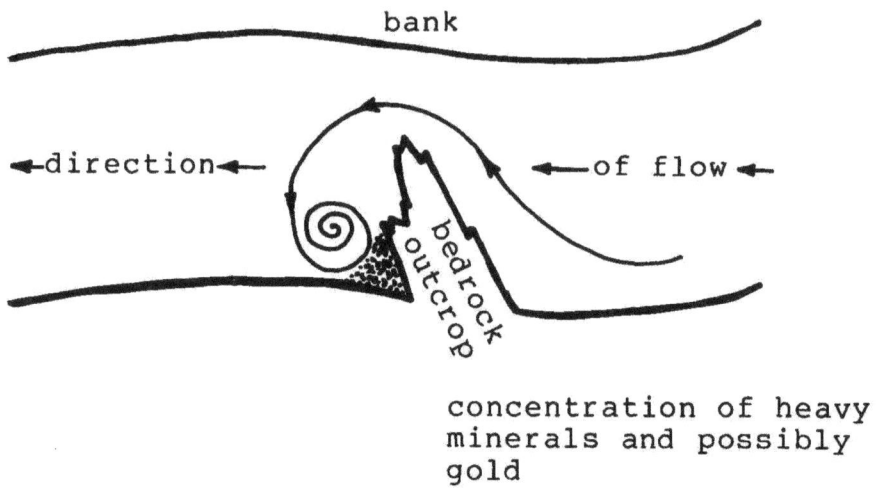

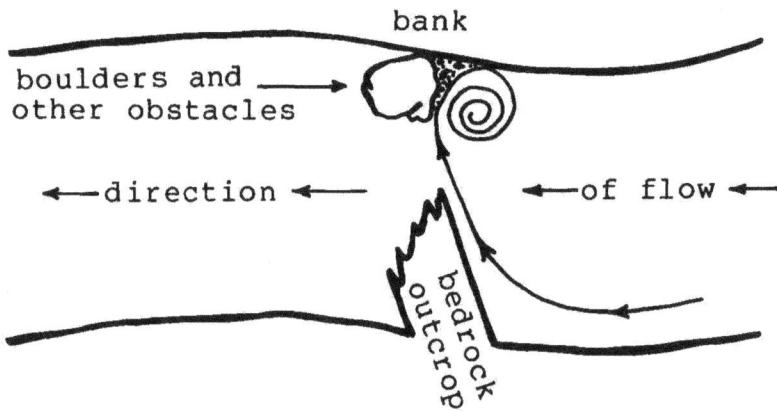

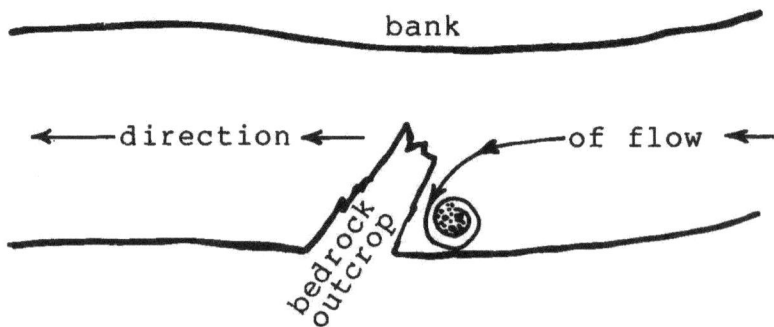

Fig. 24   Location of heavy mineral concentrations
          upstream of bedrock outcrops
          (According to Thornton, 1977)

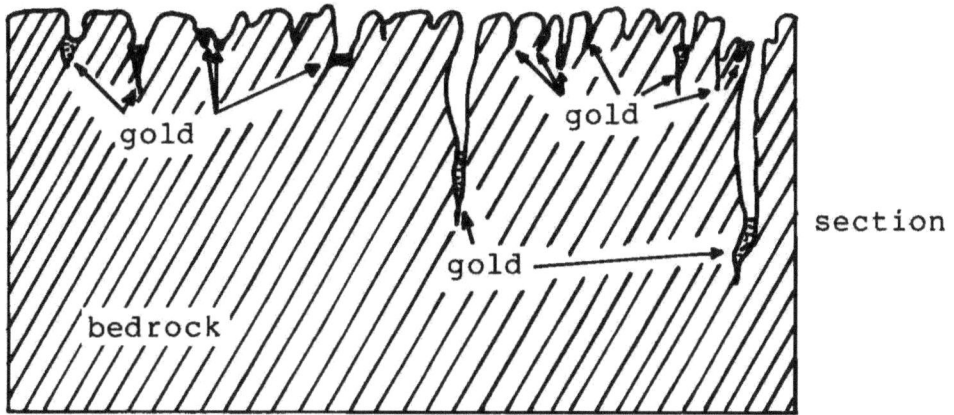

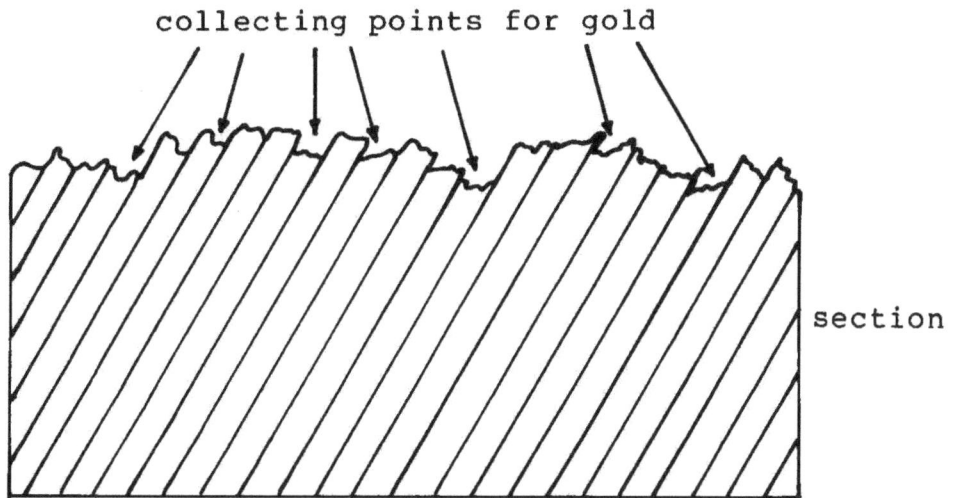

Fig. 25   Structure of the bedrock of the river bottom
           suitable for gold concentration.
           (According to Thornton, 1977)

If the gold content is very high one might have to consider damming the river upstream of the find and diverting it. The volume of sand and gravel (V) is determined by the product of the length (L) of the river which contains gold, the thickness of the sand and gravel (D) and the width of the river (B), so that:

$$V = L \times D \times B$$

The value of the deposit (W) is

$$W = V \times G \times C \times P$$

One should be aware that generally in the upper reaches of a river the overburden is thinner and the gold particles coarser, whereas in the lower reaches the overburden becomes thicker and the gold finer.

4.3.1.3. Secondary Colluvial Deposits

These are found at the foot of hills and in river valleys, below the region of primary gold bearing bedrock.

Pit Location

During exploration a series of pits should be dug a few centimeters into the weathered bedrock, starting with a line uphill from the location of the first find, as shown in Fig. 20. The first indication may have been found during prospecting, when the secondary colluvial character was recognised, or it could be a point in a river upstream of which there is a steadily increasing gold content. It is obvious that the gold was brought to this point either from the primary bedrock in the bottom of the river, or from secondary colluvial deposits on the river banks, see Fig. 5 and 26. Exploration proceeds uphill by digging pits every 25m until there is no longer any gold. This indicates that one has passed the gold bearing bedrock. The gold content of the pits and the nature of the deposit are recorded in the same way as for stream or land deposits. It will often be found difficult to distinguish between the layers of sand, gravel and clay as they may occur in confused interbeddings. If there are indications that there are several quartz reefs within a short distance

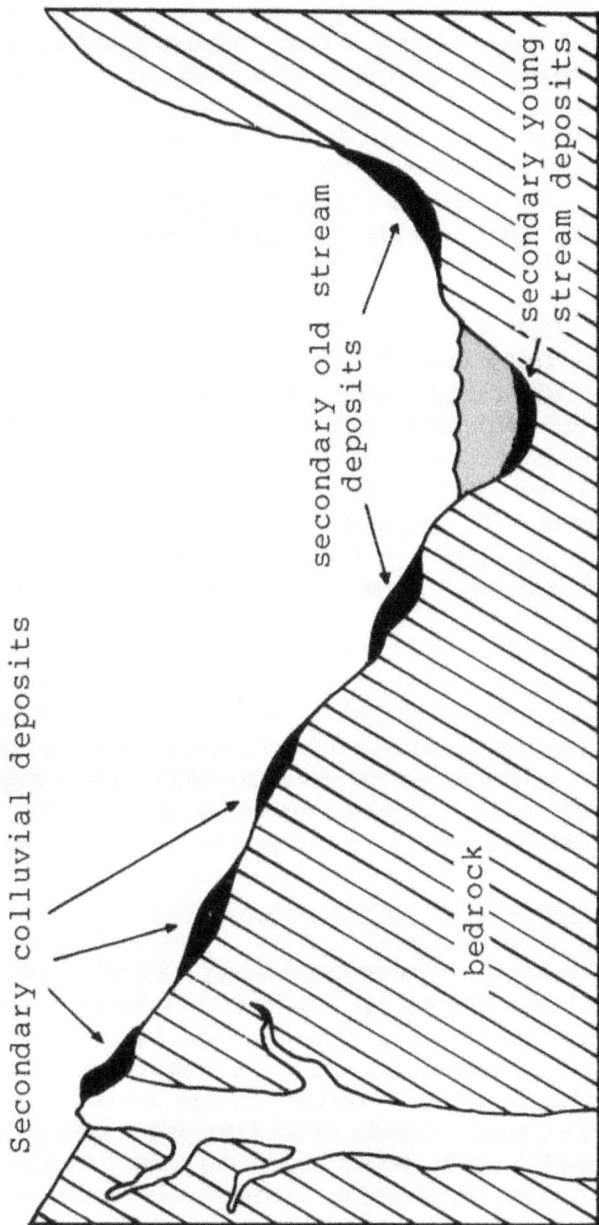

quartz reef (primary deposit)

Fig. 26 Relationship between primary and secondary
gold deposits
(According to Thornton, 1977)

-42-

of each other, it may be advisable first to dig a
trench at the foot of the hill in the direction
of the valley to determine the width and
direction of the zones. The value of the ground
is determined as for river deposits.

One problem which may arise is the availability
of water necessary for the operation of the
batea, rocker, ground sluice, or longtom. The
large quantities required for longtom or ground
sluicing can be supplied by a 1.25 or 1.50 inch
pump over distances of 50 to 75m (as shown in
Fig. 20) if there is enough water in a side
stream, or for 150m if one has to use the main
stream.

If the shape of the valley and the direction of
the gold bearing bedrock allow it, one should
consider whether the overburden and the gold-
containing deposit should be removed by monitors
(high pressure water jets). The suction dredge
can be used for this purpose, by removing the
suction hose from the vacuum chamber and fitting
a smaller pipe. A full description of this
method is beyond the scope of this manual.

Although colluvial deposits are generally smaller
than alluvial or river deposits, they can still
be attractive because of the relatively thin
overburden and the occurrence of nuggets. In
practice one starts with stream deposits and
works upstream, checking the sides for colluvial
deposits and possibly gold bearing bedrock.

4.3.2. Exploration of Primary Deposits

If during the prospecting phase of colluvial
exploration, sharp edged pieces of quartz
containing gold are found, it is recommended that
these are sampled in a systematic way to
establish the direction of the gold
mineralisation. Information can be gathered from
geochemical and prospecting if values of more
than 0.2 ppm gold are found along several lines,
which can be connected as shown in Fig. 17.

Information about the direction of mineralization
can also be established by digging pits as shown
in Fig. 21, enabling the direction to be studied
in three dimensions by observing the sides of the
pit. Once the direction of the reef or of the
series of reefs is known then this can be
followed by taking rock chip samples every 25 to

100m, depending on the nature of the terrain. This is done by chipping pieces, 2-3cm in size, off the quartz vein until a composite sample of about 2kg is obtained. Half the weight of each sample (1kg) should be put into a linen bag bearing a label, with this information:

- name of the concession holder

- registration number of the concession

- name of the concession and the distance of the sample from the first point of the line.

In the example shown in Fig. 17 the following samples were collected:-

No. 1	Line A	picket 5	0-100m
No. 2	Line A	picket 5	100-200m
No. 3	Line A	picket 5	200-300m

These samples were sent to the laboratory of the GMD for the determination of gold content and used as a check on the samples assayed in the field.

The other half of the samples (1kg) remains in the field, is broken up with a hammer and ground in a mortar until it has the appearance of sand. It is then treated in a batea and the gold weighed so that the content, which can be separated mechanically, is determined and expressed in grams per ton. In this type of field investigation, one has to bear in mind that the gold, being soft, is beaten into thin platelets by the hard quartz, so that gravitational methods can indicate a lower gold content than is actually present.

If the gold content is of the order of 20g/t then it is recommended that a trench be dug at the point of highest gold content at right angles to the direction of the reef in order to expose the hard rocks beneath the surface. In this way the sampling of the reef is improved and information regarding the width and nature of the occurrence can be derived. Generally, this operation requires the removal of comparatively large amounts of ground, and it might be advantageous to use a bull-dozer or dragline. If high gold contents are found over a distance of more than 500m over a width of more than 1m then one should investigate, by means of drilling, how the reef and its mineralization continues in depth. It will be clear that this requires considerable investment in geological and engineering aspects, taking a highly professional approach which is beyond the scope of this manual. In this case one should certainly consider a more modest exploitation of the surrounding colluvial deposits.

5. PURITY OF GOLD

Gold which has been separated by gravity methods will not be completely pure, as silver, copper and other impurities are mixed with it. The purity of gold (bullion) is expressed in parts per 1000. The average purity of placer gold (i.e. secondary and colluvial) in Suriname is 900. The samples are tested by the Assay Office of the Weights and Measures Department of the Ministry of Economics.

6. AMALGAMATION OF GOLD FROM MINERAL CONCENTRATES WITH MERCURY

    a. Extract the heavy concentrate from the original sample by means of a batea, longtom, rocker or other suitable method.

    b. Put the concentrate in a pan and count the gold particles of the various sizes and remove by hand any which are larger than 2mm.

    c. Add one drop of gold free mercury about the size of a bean and reduce the amount of heavy minerals to a smaller volume.

    d. Remove the mercury and put it into a 250ml heat resistant beaker, add 40 to 50ml dilute nitric acid and dissolve it until the size of the mercury drop is reduced to the size of a match head. Pour everything into a porcelain cup, add more acid and heat for a while until the mercury has been dissolved. The fine gold will remain as a spongy mass as the solution is separated. Boiling of the solution must be avoided.

-45-

e. Pour off the acid solution and wash
    the gold three or four times with warm
    water. Add one or two drops of
    alcohol and dry the gold residue by
    applying a little heat.

f. Heat the flameproof porcelain beaker
    until it starts to glow and all the
    remaining mercury is evaporated.

g. Put the gold aggregate together with
    the plus 2mm particles, which were
    previously separated, on to scales and
    record the weight.

The following, simpler, method was developed but
more mercury may be retained in the sample. The
fine gold, mixed with small gangue particles is
heated with mercury in a can which has been cut
in half and is covered with a leaf. The mercury
evaporates and precipitates on the leaf while the
gold remains behind as a spongy lump, which may
still contain stone particles and mercury. The
mercury can be recovered by pulverizing the leaf
in a small tin can, and can then be re-used.

7. TREATMENT OF POISONING

7.1 Irritants including Mercury

Besides mercury there are other poisons from which small scale miners may be at risk: solutions containing copper, lead, phosphorus and even normal day-to-day things such as petrol, paraffin and iodine. All are irritant poisons.

Symptoms

All these poisons irritate and inflame the stomach and intestines, causing retching, purging, colicky pains, and ultimately suffocation and collapse. In iodine poisoning there is also great thirst.

Treatment

If able to swallow, give an emetic. Dilute the poison by giving water, tea or milk in abundance. Keep the patient warm and give him beaten-up eggs or salad oil, except in the case of phosphorus poisoning. In iodine poisoning give starch and water.

NB. - Phosphorus dissolves in oil, and in this form is more readily absorbed into the system. Therefore never give oil in a case of phosphorus poisoning.

7.2 Corrosives

This group comprises all the strong acids and alkalis, many of which are in regular use in all types and size of mines. These include sulphuric acid, hydrochloric acid, nitric acid, oxalic acid, carbolic acid, creosote and alkalis such as lime, caustic soda, potash and ammonia. These poisons stain, burn and eat into the tissue of the mouth, throat, gullet and stomach, causing intense pain, suffocation and collapse.

Treatment

In general no emetic can be given, as the spasm of vomiting might rupture the wall of the stomach with very serious results if it is already corroded. For oxalic acid an emetic may be given followed by a weak alkaline solution.

In all cases dilute at once by giving water, after which for carbolic acid, creosote and lysol give 2 to 3 tablespoonfuls of Epsom Salts or Glauber Salts if available. Otherwise, and for all other corrosive acid poisons, neutralize by giving weak alkaline solutions as shown in the following rules, paragraph 4(b).

7.3   General Rules for the Treatment of Poisoning

1. Send for a doctor, stating what has occurred and the suspected poison.

2. If breathing is not perceptible, start artificial respiration immediately.

3. Preserve any traces of the poison which may exist. Carefully retain any glass, cup, bottle, packet or food which may have contained some of the poison; also the patients vomit, if any, and stained clothing. This is of legal as well as medical importance.

4. (a) If the patient is conscious, can swallow, and has no burns or blisters about the lips or mouth, promote vomiting in order to rid the system of as much poison as possible. This can be done as follows:-

   By tickling the back of the throat with a finger, a feather or a paper spill.

   By emetics, such as two teaspoonfuls of mustard or one tablespoonful of salt in half a tumbler of warm water, or two teaspoonfuls of ipecacuanha wine. Repeat the dose every five minutes until vomiting begins.

   (b) If the lips or mouth are burned, give no emetic. The poison was either a strong acid or alkali. Discover which and neutralize by administering the other. For acid poisoning give repeated doses of a weak alkali such as chalk, magnesia, whitening, or ceiling plaster. For alkaline poisoning, give, by the tumblerful, a weak acid such as vinegar, lime or lemon juice in an equal quantity of water. If it is not known what the poison was give copious draughts of cold water.

5. Administer the proper antidote. This will counteract the poison and render it harmless. For acids and alkalis see paragraph 4 above.

If there is no clue as to the nature of the poison, give milk, raw eggs beaten up in either milk or water, beaten-up cream and flour, strong tea or even just water alone.

6. Treat any special symptoms.

   (a) For shock and collapse promote warmth and give stimulants.

   (b) For drowsiness, keep the patient on the move.

   (c) If the throat is badly swollen, apply hot fomentations to reduce the risk of the air passages becoming blocked and give frequent sips of cold water.

8.   REFERENCES

Anderson, R. B., 'The Exploration, Valuation and Development of Small Mines', Proceedings of a symposium arranged by the Institution of Mining and Metallurgy, Rhodesia Section, 1965, 104pp.

Armstrong, A. T., Handbook on Small Mines. (Eastwood) South Australian Dept. of Mines and Energy, 1980, 206pp

Bernewitz, M. W. Von, A.B.C. of Practical Placer Mining, Great Western Publishing Co., 62pp.

Bernetwitz, M. W. von, 'Saving Gold By Means of Corduroy', 1939, U.S. Bureau Mines Inf. Circ 7085, 17pp.

Boericke, W. F., Prospecting and Operating Small Gold Placers, New York: Wiley, 1936, 144pp,

Clifton, Hunter, Swanson and Phillips, 'Sample Size and Meaningful Gold Analysis', 1969, U.S. Geol. Survey Prof. Paper 625-C.

Fisher, R.P., and Fischer, F.S., 'Interpreting Pan Concentrate Analysis of Stream Sediments in Geochemical Exploration for Gold', 1970, U.S. Geol. Survey Circ. 592, 9pp.

Francis, T.G., 'So you want to do some mining (Alluvial Mining of Gold and Tin: Hints on Equipment used)', 1977, Brisbane: Queensland Govt. Min. Journal 78/914, p.601-604.

Gold Extraction for the Small Operator, 2nd Ed. 103pp (London- Imperial Chemical Industries Ltd., 1943.)

Griffiths, S.V., Alluvial Prospecting and Mining, 1960, New York: Pergamon Press Inc., 245pp.

Haffty, Riley and Goss, A Manual of Fire Assaying and Determination of Noble Metals in Geological Materials, 1977, U.S. Geol, Survey Bull. 1445, 58pp.

Harrison, H.L.H., Valuation of Alluvial Deposits, London: Mining Publications Ltd., 1954, 308pp.

-50-

Harrison, H.L.H., Alluvial Mining for Tin and Gold, London: Mining Publications Ltd., 1962, 313pp.

Hawkes, H.E., 'Principles of Geochemical Prospecting', 1957, U.S. Geol. Survey Bull. 1000-F.

Idriess, I.L., Prospecting for Gold, New York: McGraw-Hill Co., 359pp.

Lock, C.G.W., Practical Gold-Mining, (London: E & F N Spon, 1889) 788pp, figs., photos., refs.

McGowan, A., Plans for Building a Portable Gold Rocker. United Prospectors, 166 West High Street, Benicia, CA 94510,

McGowan, A., The New Method of Gold Mining, Gold Rush, P. O. Box 1882, Newport Reach, CA 94510,

McGowan, A., The Extraction of Free Gold, 1973, Carson Enterprises, 801 Juniper Avenue, Boulder, CO 80302.

McLellan, Berkenkotter, Wilmont and Stahl, Drilling and Sampling Tertiary Gold Placer in Nevada County, California, U.S. Bureau Mines Rept. Inv. 7935, 50pp.

Moen W.S., and Hunting, M.T., Handbook for Gold Prospectors in Washing, 1975, Washington Dept. Nat. Resources, Div. Geol. and Earth Resources Inform. Circ. 57.

Mineral Information Service of California, Elementary Placer Mining Methods, San Francisco, 10, Aug. 1 1957, 1-7.

Mineral Information Service of California, Prospecting, Exploring and Developing the Small Mine, San Francisco, 11, Dec. 1958, 1-6.

Mines Geol., Special Publications Division, Basic Placer Mining, No. 41 16pp.

Montana School of Mines, 'Placer Mining Workshop', Butte, June 29-30, 1974.

Taylor, Prisbrey, Green and Hoskins, The Design, Economics, Mining and Metallurgy of Small Scale Gold and Silver Recovery Operations, 1980, Dept. of Mining and Metallurgy, College of Mines, Univ. Idaho, Moscow, Idaho.

Powers, T. W., Economic Processes for the Recovery of Gold and Silver, 1980, Dept. of Mining and Metallurgy, College of Mines, Univ. Idaho, Moscow, Idaho.

Powers, T.W., Tabling Gold Ores. T. Powers, P. O. Box 585, Hilmar, CA 95324,

Raeburn, C., and Milner, H.B., Alluvial Prospecting: The Technical Investigation of Economic Alluvial Minerals. London: Thos. Murby & Co., and New York: D. van Nostrand Co., 1927, 478pp.

Wells, J.H., Placer Examination, Principles and Practice, 1973, Bur. Land Management Tech. Bull. 4, 209pp.

West, J., How to Mine and Prospect for Placer Gold, 1971, U.S. Bureau Mines Inf. Circ. 8517.

Wolff, E., Handbook for the Alaskan Prospector, 1969, Mineral Ind. Research Lab., Univ. Alaska.

www.ingramcontent.com/pod-product-compliance
Lightning Source LLC
Chambersburg PA
CBHW042356030426
42336CB00030B/3500